Theory of Copper Oxide Superconductors

Hiroshi Kamimura   Hideki Ushio
Shunichi Matsuno   Tsuyoshi Hamada

Theory of Copper Oxide Superconductors

With 109 Figures

 Springer

Prof. Dr. Hiroshi Kamimura
Tokyo University of Science
Department of Applied Physics
1-3 Kagurazaka
Shinjuku-ku
Tokyo 162-8601, Japan
e-mail: kamimura@rs.kagu.tus.ac.jp

Dr. Shunichi Matsuno
Tokai University
General Education Program Center
Shimizu Campus
3-20-1 Shimizu-Orido
Shizuoka 424-8610, Japan
e-mail: smatsuno@scc.u-tokai.ac.jp

Prof. Dr. Hideki Ushio
Tokyo National College of Technology
1220-2 Kunugida-chou
Hachioji 193-0997, Japan
e-mail: ushio@tokyo-ct.ac.jp

Dr. Tsuyoshi Hamada
Mizuho Information and Research Institute
2-3 Kanda-Nishiki-cho
Chiyoda-ku
Tokyo 101-8443, Japan
e-mail: tsuyoshi.hamada@gene.mizuho-ir.co.jp

Library of Congress Control Number: 2005924330

ISBN-10 3-540-25189-8 Springer Berlin Heidelberg New York
ISBN-13 978-3-540-25189-7 Springer Berlin Heidelberg New York

This work is subject to copyright. All rights are reserved, whether the whole or part of the material is concerned, specifically the rights of translation, reprinting, reuse of illustrations, recitation, broadcasting, reproduction on microfilm or in any other way, and storage in data banks. Duplication of this publication or parts thereof is permitted only under the provisions of the German Copyright Law of September 9, 1965, in its current version, and permission for use must always be obtained from Springer. Violations are liable for prosecution under the German Copyright Law.

Springer is a part of Springer Science+Business Media
springeronline.com

© Springer-Verlag Berlin Heidelberg 2005
Printed in Germany

The use of general descriptive names, registered names, trademarks, etc. in this publication does not imply, even in the absence of a specific statement, that such names are exempt from the relevant protective laws and regulations and therefore free for general use.

Typesetting: by the authors and TechBooks using a Springer LATEX macro package
Cover design: *design & production* GmbH, Heidelberg

Printed on acid-free paper   SPIN: 10950906   57/3141/jvg   5 4 3 2 1 0

Preface

The objective of this book is to provide an up-to-date comprehensive description of the Kamimura–Suwa model, which is the first of the present representative two-component theories in high temperature superconductivity. In 1986 George Bednorz and Karl Alex Müller made the remarkable discovery of superconductivity with an unbelievingly high value of $T_c = 35$K, by substituting Ba^{2+} ions for La^{3+} ions in the antiferromagnetic insulator La_2CuO_4. Soon after this discovery T_c rose to 90 K by synthesizing $YBa_2Cu_3O_{7-\eta}$ with a deficit in oxygen. Further exploration for new copper oxide superconducting materials with higher T_c led to the discovery of Bi–Sr–Ca–Cu–O, Tl–Ba–Ca–Cu–O and Hg–Ba–Ca–Cu–O compounds in subsequent years. The new class of copper oxide compounds mentioned above is called "cuprates". At present $T_c = 135$ K under ambient pressure and $T_c = 164$ K under 31 GPa observed in $HgBa_2Ca_2Cu_3O_8$ are the highest value so far obtained. The Kamimura–Suwa model, which was originally developed in 1993, is a theory of these real copper oxide superconducting materials. Since undoped La_2CuO_4 is a Mott–Hubbard antiferromagnetic insulator, its electronic structure can not be explained by the ordinary one-electron energy band theory. In this context the important role of electron-correlation was pointed out.

On the other hand, a d-hole state in each Cu^{2+} ion in the ligand field with octahedral symmetry is orbitally doubly-degenerate so that it is subject to strong Jahn–Teller interaction in La_2CuO_4. As a result, a CuO_6 octahedron in La_2CuO_4 is elongated along the c-axis due to the Jahn–Teller distortion. In this circumstance most proposed models so far assume that hole-carriers itinerate in the CuO_2 layers perpendicular to the c-axis. However, when hole-carriers are doped into cuprates, CuO_6 octahedrons or CuO_5 pyramids are deformed so as to minimize the total electrostatic energy of a whole system. We call this kind of deformation the "anti-Jahn–Teller effect", because the CuO_6 octahedrons or CuO_5 pyramids elongated by the Jahn–Teller interaction along the c-axis in undoped materials shrink along the c-axis so as to partly cancel the energy gain due to the Jahn–Teller effect by doping the carriers. As a result, the energies of two kinds of orbital become closer again.

The Kamimura–Suwa model, abbreviated as the K–S model, takes account of both effects of the electron correlation and lattice distortion due to the anti-Jahn–Teller effect on equal footing. As a result, two kinds of multiplet in the presence of local antiferromagnetic order due to the localized spins play an important role in determining the electronic structures of cuprates and creating the d-wave pairing mechanism of superconductivity. This book clarifies the important roles of both electron correlation and lattice distortion in real cuprate materials with hole-doping. In particular, it is clarified that two-component scenario and inhomogeneity are key factors in high temperature superconductivity in cuprates. Eleven chapters among the 14 chapters in this book are devoted to describing the many-body-effect-including electronic structures, Fermi surfaces, normal-state properties of superconducting cuprates and the mechanism of high temperature superconductivity in an instructive way, based on the K–S model. Readers will understand that a number of theoretical predictions by the K–S model have been proven experimentally by various recent experiments. This book is written in a self-contained manner in which, for the most part, readers will understand the basic physical foundations even if they are not trained in advanced many-body techniques.

Tokyo
March 2005

Hiroshi Kamimura
Hideki Ushio
Shunichi Matsuno
Tsuyoshi Hamada

Contents

1. Introduction ... 1
2. **Experimental Results of High Temperature Superconducting Cuprates** ... 9
   - 2.1 Introduction ... 9
     - 2.1.1 Basic Crystalline Structures of Cuprates ... 9
     - 2.1.2 Ability of Changing Carrier Concentration ... 11
   - 2.2 Experimental Results of Cuprates ... 11
     - 2.2.1 The Phase Diagram of Cuprates ... 12
     - 2.2.2 The Symmetry of the Gap ... 13
3. **Brief Review of Models of High-Temperature Superconducting Cuprates** ... 15
   - 3.1 Introduction ... 15
   - 3.2 Brief Review of Theories for HTSC ... 17
     - 3.2.1 Jahn–Teller Polarons and Bipolarons ... 17
     - 3.2.2 The Resonating Valence Bond (RVB) State and Quasi-particle Excitations ... 18
     - 3.2.3 The d–p Model ... 21
     - 3.2.4 The t–J Model ... 22
     - 3.2.5 Spin Fluctuation Models ... 24
     - 3.2.6 The Kamimura–Suwa Model and Related Two-Component Mechanisms ... 25
4. **Cluster Models for Hole-Doped CuO_6 Octahedron and CuO_5 Pyramid** ... 29
   - 4.1 Ligand Field Theory for the Electronic Structures of a Single Cu^{2+} Ion in a CuO_6 Octahedron ... 29
   - 4.2 Electronic Structures of a Hole-Doped CuO_6 Octahedron ... 30
   - 4.3 Electronic Structure of a Hole-Doped CuO_5 Pyramid ... 30
   - 4.4 Anti-Jahn–Teller Effect ... 31
   - 4.5 Cluster Models and the Local Distortion of a Cluster by Doping Carriers ... 33

5 MCSCF-CI Method: Its Application to a CuO_6 Octahedron Embedded in LSCO 37
5.1 Description of the Method 37
5.2 Choice of Basis Sets in the MCSCF-CI Calculations 38
5.3 Calculated Results of Hole-Doped CuO_6 Octahedrons in LSCO 39
   5.3.1 The $^1A_{1g}$ Multiplet (the Zhang–Rice Singlet) 39
   5.3.2 The $^3B_{1g}$ Multiplet (the Hund's Coupling Triplet) 40
5.4 Energy Difference between Zhang–Rice Singlet ($^1A_{1g}$) and Hund's Coupling Triplet ($^3B_{1g}$) Multiplets 41

6 Calculated Results of a Hole-Doped CuO_5 Pyramid in $YBa_2Cu_3O_{7-\delta}$ 43
6.1 Introduction 43
6.2 Energy Difference between $^1A_1$ and $^1B_1$ Multiplets 44
6.3 Effect of Change Density Wave (CDW) in a Cu–O Chain 45

7 Electronic Structure of a CuO_5 Pyramid in $Bi_2Sr_2CaCu_2O_{8+\delta}$ 51
7.1 Introduction 51
7.2 Models for Calculations 51
7.3 Calculated Results 52
7.4 Remarks on Cuprates in which the Cu–Apical O Distance is Large 53

8 The Kamimura–Suwa (K–S) Model: Electronic Structure of Underdoped Cuprates 55
8.1 Description of the Model 55
8.2 Experimental Evidence in Support of the K–S Model 58
   8.2.1 Existence of the Antiferromagnetic Spin Correlation in the Underdoped Regime 58
   8.2.2 Coexistence of the $^1A_{1g}$ and the $^3B_{1g}$ Multiplets 58
8.3 Hamiltonian for the Kamimura–Suwa Model (The K–S Hamiltonian) 59
8.4 Concluding Remarks 61

9 Exact Diagonalization Method to Solve the K–S Hamiltonian 63
9.1 Introduction 63
9.2 Description of the Method: Lanczos Method 63
9.3 Calculated Results for the Spin-Correlation Functions 66
9.4 Calculated Results for the Orbital Correlation Functions 74

	9.5 The Case of a Single Orbital State	76
	9.6 Calculated Results of the Radial Distribution Function for Two Hole-Carriers	79
10	**Mean-Field Approximation for the K–S Hamiltonian**	**83**
	10.1 Introduction	83
	10.2 Slater–Koster Method: Its Application to LSCO	85
	10.3 Computation Method to Calculate the Many-Electron Energy Bands: Its Application to LSCO	87
	10.4 Computation Method Applied to YBCO Materials	92
	10.5 Appendix	98
11	**Calculated Results of Many-Electrons Band Structures and Fermi Surfaces**	**107**
	11.1 Introduction	107
	11.2 Calculated Band Structure Including the Exchange Interaction between the Spins of Hole-Carriers and Localized Holes	107
	11.3 Calculated Fermi Surface and Comparison with Experiments	110
	11.4 Wavefunctions of a Hole-Carrier with Particular k Vectors and the Tight Binding (TB) Functional Form of the #1 Conduction Band	115
	11.5 Calculated Density of States	117
	11.6 Remarks on the Simple Folding of the Fermi Surface into the AF Brillouin Zone	118
12	**Normal State Properties of $La_{2-x}Sr_xCuO_4$**	**121**
	12.1 Introduction	121
	12.2 Resistivity	122
	12.3 Hall Effect	125
	12.4 Electronic Entropy	127
	12.5 Validity of the K–S Model in the Overdoped Region and Magnetic Properties	130
	12.6 The Origin of the High-Energy Pseudogap	131
	12.6.1 Introduction	131
	12.6.2 Calculation of Free Energies of the SF- and LF-Phases	132
	12.6.3 Origin of the "High-Energy" Pseudogap	135
13	**Electron–Phonon Interaction and Electron–Phonon Spectral Functions**	**139**
	13.1 Introduction	139
	13.2 Calculation of the Electron–Phonon Coupling Constants for the Phonon Modes in LSCO	139

| | 13.3 Calculation of the Spectral Functions
for s-, p- and d-waves .... 143 |
| --- | --- |
| | 13.4 Appendix .... 154 |

14 Mechanism of High Temperature Superconductivity 161
14.1 Introduction 162
14.2 Appearance of Repulsive Phonon-Exchange Interaction
in the K–S Model 163
    14.2.1 The Selection Rule 163
    14.2.2 Occurrence of the d-wave Symmetry 167
14.3 Suppression of Superconductivity by Finiteness
of the Anti-Ferromagnetic Correlation Length 171
    14.3.1 Influence of the Lack
of the Static, Long-Ranged AF-Order
on the Electronic Structure 172
    14.3.2 Suppression of Superconductivity
Due to the Finite Lifetime Effect 173
14.4 Strong Coupling Treatment
of Conventional Superconducting System 174
    14.4.1 Green's Function Method in the Normal State 174
    14.4.2 Application of the Green's Function Method
to a Superconducting State 179
    14.4.3 Inclusion of Coulomb Repulsion 181
    14.4.4 T_c-Equation in the Strong Coupling Model 181
14.5 Application of McMillan's Method
to the K–S Model 183
14.6 Calculated Results
of the Superconducting Transition Temperature
and the Isotope Effects 187
    14.6.1 Introduction 187
    14.6.2 The Hole-Concentration Dependence of T_c 189
    14.6.3 Isotope Effects 191
14.7 Final Remarks 192

References 193

Subject Index 203

Index 205

1 Introduction

Superconductivity was first discovered in Hg at 4.2 K in 1911 by Heike Kamerlingh Onnes [1]. Since then, superconductivity has been found in many metallic elements of the periodic systems, alloys and intermetallic compounds. Theory of superconductivity was given by Bardeen, Cooper and Schrieffer (BCS) in 1957 [2]. Until 1986, T_c of 23.4 K observed in Nb_3Ge was the highest. In 1986 George Bednorz and Karl Alex Müller [3] made the remarkable discovery of superconductivity that brought an entirely new class of solids with an unbelievably high value of $T_c = 35$ K to the world of physics and materials science. A new superconducting material was La_2CuO_4 in which the ions of Ba^{2+}, Sr^{2-} or Ca^{2+} were doped to replace some of La^{3+} ions and hole-carriers are created. This discovery was the dawn of the era of high temperature superconductivity(HTSC). Soon after this discovery, T_c rose to 90 K when La^{3+} ions were replaced by Y^{3+} ions and a superconducting material of $YBa_2Cu_3O_{7-\eta}$ with a deficit in oxygen was made [4]. Further exploration for new copper oxide superconducting materials with higher T_c has led to the discovery of Bi–Sr–Ca–Cu–O [5], Tl–Ba–Ca–Cu–O [6] and Hg–Ba–Ca–Cu–O [6, 7] compounds in subsequent several years. At present $T_c = 135$ K under ambient pressure and $T_c = 164$ K under 31 GPa observed in $HgBa_2Ca_2Cu_3O_8$[8] are the highest values so far obtained. The historical development of superconducting critical temperature since 1911 is schematically shown in Fig. 1.1. The new class of copper oxide compounds mentioned above are called "cuprates".

The crystal structure of cuprates are layered-perovskite, consisting of CuO_2 planar sheets and interstitial insulating layers. Since the latter layers block the CuO_2 interlayer interactions, those are called "blocking layers".

In order to help the readers of this book with regard to the overall understanding of the crystal structures of cuprates at the beginning, we sketch the features of the crystal structures of representative superconducting hole-doped cuprates from $(La, Ba)_2CuO_4$ with $T_c = 35$ K to $TlBa_2Ca_2Cu_3O_9$ with $T_c = 115$ K ($HgBa_2Ca_2Cu_3O_8$ with $T_c = 135$ K) in Fig. 1.2, although we describe the crystal structures of cuprates in detail in the following chapter. The crystal structure of cuprates is classified into three types according to the types of Cu–O networks, such as octahedron type (T-phase), square-type (T'-phase) and pyramid-type (T^*-phase) [9]. They are also classified

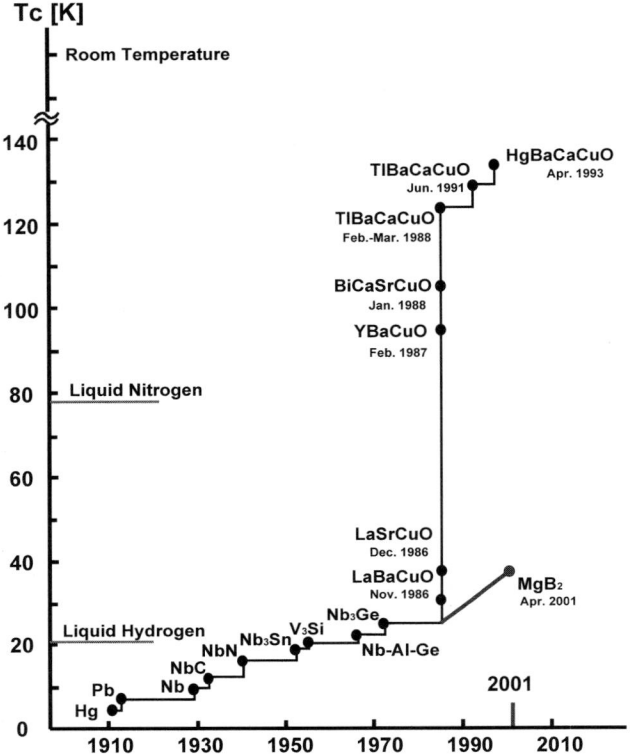

Fig. 1.1. The historical development of critical temperatures in the superconducting materials

by the number of CuO_2 sheets. In the case of lanthanum compounds, hole-carriers are produced by substituting Ba^{2+}, Sr^{2+} or Ca^{2+} ions for La^{3+} ions in La_2CuO_4, while in other cuprates hole-carriers are provided from excess oxygen in blocking layers. These hole carriers mainly move along the CuO_2 planes.

The parent compound La_2CuO_4 is experimentally an antiferromagnetic (AF) insulator[10] with a Néel temperature $T_N = 240\,\mathrm{K}$ for three-dimensional AF ordering. When an ordinary one-electron band theory is applied to the electronic structure of La_2CuO_4, it gives a metallic state, because each Cu^{2+} ion in La_2CuO_4 has one d hole. Thus one-electron band calculations are not applicable to La_2CuO_4. Besides this fact, there is a good deal of evidence that even the normal state properties of the cuprates differ remarkably from those of ordinary metals and superconductors and that conventional one-electron band theory may not be applicable.

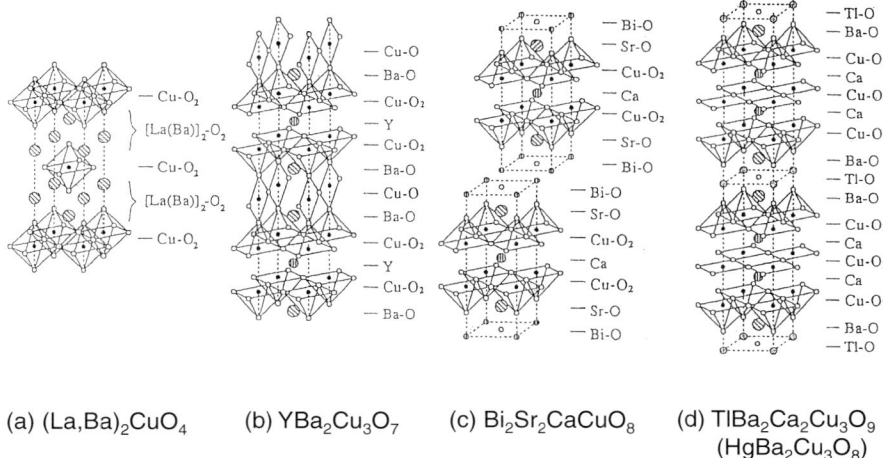

Fig. 1.2. Sketch of crystal structures of representative hole-doped cuprates

In the context of failure of the application of one-electron band theory to explain the experimental fact that La_2CuO_4 is an antiferromagnetic insulator, the important role of strong electron correlation was pointed out first by Phillip Anderson [11, 12]. On the other hand, Bednorz and Müller pointed out the important role of the strong electron-lattice interactions in cuprates, from the standpoint that a d-hole state in each Cu^{2+} ion in La_2CuO_4 is orbitally doubly-degenerate so that it is subject to strong Jahn–Teller interaction [1, 13]. In fact, a CuO_6 octahedron in La_2CuO_4 is elongated along the c-axis due to the Jahn–Teller distortion and a d hole in each Cu^{2+} ion occupies a $d_{x^2-y^2}$ orbital, where the z-axis is taken along the c-axis [14]. Since the discovery of high temperature superconductivity, many theories, including conventional and unconventional mechanisms, have been proposed.

In order to clarify the mechanism of high temperature superconductivity, however, we first have to know the electronic structures of cuprates. In this respect there seems to be general agreement that CuO_2 layers play a main role in superconductivity and normal-state transport. In this view, most theories proposed so far are based on an orbital consisting of Cu $d_{x^2-y^2}$ orbital and O p_σ orbital extended over a CuO_2 layer. However, the hole-doped cuprates consist of pyramid-type CuO_5 or octahedron-type CuO_6 clusters and have shown different features of superconducting and normal-state properties experimentally. Thus it seems difficult to explain the different features of superconducting and normal-state properties of various cuprates by using a theory based on CuO_2 layers.

Furthermore, when hole-carriers are doped into cuprates by the substitution of divalent ions for lanthanum trivalent ions in lanthanum cuprate compounds or by the introduction of excess or deficit of oxygen, CuO_6 octahedron clusters or CuO_5 pyramid clusters are deformed so as to minimize the total electrostatic energy of a whole system. We call this kind of deformation "anti-Jahn–Teller effect", because the CuO_6 octahedrons or CuO_5 pyramids elongated by the Jahn–Teller interaction along the c-axis in undoped materials are deformed so as to partly cancel the energy gain due to the Jahn–Teller effect by doping the carriers. In the case of $La_{2-x}Sr_xCuO_4$ (abbreviated as LSCO hereafter), for example, apical oxygen in an elongated CuO_6 octahedron along the c-axis in La_2CuO_4 tends to approach Cu ions by the "anti-Jahn–Teller effect". Thus, in LSCO, elongated and contracted octahedrons are mixed. This does not mean that, in the underdoped regime of low hole-concentration, only ten to fifteen percent of CuO_6 octahedrons are deformed by the anti-Jahn–Teller effect while the remaining octahedrons are elongated. In order to reduce the kinetic energy of hole-carriers, the hole-carriers may be considered to move in an averagely deformed crystal. Thus a hole-carrier may have a character of a large polaron, as Müller first pointed out [1, 13].

In this context, both effects of the electron correlation and lattice distortion due to the anti-Jahn–Teller effect play important roles in determining the electronic structures of cuprates. However, most theories so far proposed mainly consider the former effect. Noticing the importance of the effects of the electron correlation and lattice distortion, Kamimura and Suwa [15] considered both effects on equal footing and developed a theory which is applicable to real cuprates. The theory by Kamimura and Suwa is now called "the Kamimura–Suwa model", which is abbreviated as "the K–S model".

The aim of this book is to clarify the important roles of both electron correlation and lattice distortion in real cuprate materials with hole-doping, based on the K–S model. Since theoretical models based on two-dimensional CuO_2 planes have been reviewed or developed by a number of review articles or texts [16, 17, 18, 19, 20, 21, 22, 114], in this book we will concentrate first on describing the electronic structures of cuprates calculated by first-principles calculations based on the K–S model. Then we will focus on applying the K–S model to calculating various physical properties of normal and superconducting states of cuprates and on investigating whether the K–S model can clarify various anomalous behaviours observed in cuprates, by comparing the calculated results with experimental results.

Although we do not intend to review the theoretical models so far proposed, we will briefly review some of important models in Chap. 3 from our personal views. From Chaps. 4 to 14 our descriptions are concentrated on the whole activity of the K–S model for hole-doping cuprates. In this book the topic of electron-doped cuprates is not included. In cuprates the CuO_6 octahedrons or CuO_5 pyramids form a CuO_2 plane and various kinds of stacking of the CuO_2 planes compose a different kind of cuprates. When hole-carriers are doped, carriers move primarily on a CuO_2 plane. In the hole-concentration of the underdoped regime, only about ten to fifteen percent of CuO_6 octahedrons or CuO_5 pyramids are occupied by holes as an average. Thus a dopant hole hops between the highest occupied states in many-electron states called "multiplets" of a CuO_6 octahedron or a CuO_5 pyramid. Thus one must calculate multiplets as accurate as possible by first-principles methods. For this purpose the method of first-principles variational calculations called multi-configuration selfconsitent field method with configuration interaction (abbreviated as MCSCF-CI method) is developed for a CuO_6 octahedron or a CuO_5 pyramid cluster with one dopant hole in Chap. 5. Before that, the description of a cluster model is given in Chap. 4. It is clear that the strong electron correlation creates localized spins around Cu sites while the anti-Jahn–Teller effect leads to the coexistence of the Zhang–Rice spin-singlet multiplet and the Hund's coupling spin-triplet multiplet, where the energies of both multiplets are nearly the same in the underdoped regime. The results of the lowest electronic states calculated by the MCSCF-CI method for LSCO, $YBa_2Cu_3O_{7-\delta}$ (abbreviated as $YBCO_{7-\delta}$) and $Bi_2Sr_2CaCu_2O_{8+\delta}$ (abbreviated as Bi2212) are presented in Chaps. 5 to 7.

In the underdoped regime, the doping concentration of hole-carriers is low. In this case, it is shown on the basis of the results of the first-principles cluster calculations that, when the localized spins form antiferromagnetic (AF) ordering in a spin-correlated region of finite size, a hole-carrier can hop between the highest occupied levels in the Zhang–Rice spin-singlet multiplet and the Hund's coupling spin-triplet multiplet without destroying the AF order. This is the essence of the Kamimura–Suwa model (K–S model). In Chaps. 8 to 10 the essence of the K–S model, the validity of the K–S model in real cuprates and an approximation method of solving the K–S Hamiltonian to represent the K–S model are described. In Chap. 11 the calculated results of the energy bands, the Fermi surfaces and the density of states are presented. In these results the many-body effects such as the exchange interactions between the carrier spins and the localized spins, etc., are included in the electronic structures of a hole-carrier system in the sense of the mean field approximation, while the localized spins form the antiferromagnetic order in the spin-correlated region. These results give an antiferromagnetic insulator when the cuprates are undoped. From this result one can understand that the energy band and Fermi surfaces calculated by the K–S model are completely different from those obtained from the ordinary Fermi-liquid picture. In Chap. 11 the calculated Fermi surfaces of LSCO are compared with the Fermi arcs observed by angle resolved photoemission experiments (ARPES). One can see a very good quantitative agreement between theory and experiment by Yoshida and his coworkers [23, 24, 25], indicating the strong support of the K–S model from experiments.

By using the results of the many-body-effect-including energy bands, Fermi surfaces and the density of states calculated by the mean-field approximation for the exchange interaction between the carrier's and localized spins, the normal-state properties such as the electrical resistivity, the Hall effect, the electronic entropy, the magnetic properties, etc., are calculated.

These calculated results are compared with experimental results of anomalous normal state properties in Chap. 12. In particular, it is clarified for the first time from the quantitative standpoint that the effective mass of a hole-carrier is about six times heavier than the free electron mass even in the well-overdoped region. According to the K–S model, an origin for the heavy mass is due to the interplay between the electron correlation and the local lattice distortion. In this respect we can say that the electron-lattice interactions in cuprates are strong and that a hole-carrier in the hole-doped cuprates has a nature of a large polaron.

In order to investigate whether the electron-lattice interaction is really strong or not, the electron-lattice interactions in cuprates are calculated for LSCO in Chap. 13, and it is shown that they are really strong. By using these calculated results, the momentum- and frequency-dependent electron–phonon spectral functions are calculated for all the phonon modes in LSCO. It is shown that the momentum-dependent spectral functions have a feature of d-wave symmetry and that the occurrence of d-wave, even for phonon-involved mechanisms, is due to the interplay between the electron–phonon interaction and the underlying AF order in the metallic state in the K–S model. Finally in Chap. 14 we first prove rigorously that a characteristic electronic structure of the K–S model causes an anomalous effective electron-electron interaction between holes with different spins and that this effective electron-electron interaction leads to d-wave symmetry in the superconducting gap function. Indeed Kamimura, Matsuno, Suwa and Ushio showed for the first time on the basis of the K–S model that the symmetry of superconductivity gap is d-wave even for the phonon-mechanism, when a superconducting state coexists with the local AF order [26, 27, 28, 29, 30].

Then the hole-concentration dependence of the superconducting transition temperature T_c and of the isotope effect α are calculated by solving the linearized Eliashberg equation. In the K–S model a metallic region in the underdoped regime is finite due to a finite spin-correlation length. However, this metallic region is more expanded in its area from the spin-correlated region by the spin-fluctuation effect characteristic of the two-dimensional Heisenberg spins systems in the AF localized spins. By determining the size of the metallic region so as to reproduce $T_c = 40$ K at the optimum doping in LSCO, one can see that the size of a metallic region at the optimum doping is about 30 nm. From this result it is concluded that the superconducting regions in the underdoped regime of hole-concentration are inhomogeneous. In this way Chap. 14 is devoted to the description of a theory of high temperature superconductivity based on the K–S model.

Acknowledgments

Our work described in this book is the happy outcome of collaboration at some stage or other with Yuji Suwa, Akihiro Sano, Kazushi Nomura,

Yoshikata Tobita, Mikio Eto, Kenji Shiraishi, Nobuyuki Shima, Atsushi Oshiyama, Takashi Nakayama, Tatsuo Schimizu, and Hideo Aoki. We would like also to thank all those people with whom we have had discussions on the subject matter of this book. In particular, Prof. Karl Alex Müller, Prof. Wei Yao Liang, Dr. John Loram, Prof. Elias Burstein, Prof. Marvin Cohen, Dr. Annette Bussmann-Holder, Prof. Takeshi Egami, Prof. William Little, Prof. Sasha Alexandrov, Prof. Juergen Haase, Dr. Alan Bishop, Dr. Steven Conradson, Prof. Robert Birgeneau, Prof. Marc Kastner, Dr. Gabriel Aeppli, Prof. Kazuyoshi Yamada, Dr. Chul-Ho Lee, Dr. C. T. Chen, Prof. Atsushi Fujimori, Prof. Takashi Mizokawa, Dr. Teppei Yoshida, Prof. Zhi- Xun Shen, Prof. Alessandra Lanzara, Dr. Hiroyuki Oyanangi, Prof. Antonio Bianconi, Dr. Naurong Saini, Prof. Masashi Tachiki, Prof. Yasutami Takada, Prof. Kazuko Mochizuki, Prof. Naoshi Suzuki, Prof. Noriaki Hamada, Prof. Chikara Ishii, Prof. Nobuo Tsuda, Prof. Nobuaki Miyakawa, Prof. Migaku Oda, Prof. Seiji Miyashita, Prof. Yoshio Kitaoka, Prof. Kazuhiko Kuroki,. Prof. Yoshio Kuramoto, Prof. Hiromichi Ebisawa, Prof. Fusayoshi Ohkawa and Prof. Naoto Nagaosa. Finally, one of the authors (H.K.) acknowledges partial financial supports from "High-Tech Research Centre" Project for Private Universities: matching fund subsidy from Ministry of Education, Culture, Sports, Science and Technology through the Frontier Research Center for Computational Sciences of Tokyo University of Science and also from CREST of Japan Science and Technology.

2 Experimental Results of High Temperature Superconducting Cuprates

2.1 Introduction

In this chapter we describe some important features of copper oxides observed from experiments. Since the discovery of high temperature superconducting cuprates, numerous experiments have been performed, and hence it is impossible to mention all of them in this book. Thus let us concentrate on experimental results which are closely related to the topics of this book.

In the following, we first show typical crystalline structures of cuprate families and point out characteristics of these structures. Then, in the following sections, various physical quantities are briefly explained.

2.1.1 Basic Crystalline Structures of Cuprates

All the cuprate families are characterized by the fact that they are all metal oxide (MO) materials with quasi-two dimensional layered structures in which copper oxide layers are always contained. They all consist of alternate stacking of copper oxide-layers (CuO_2-layers) and the so-called blocking layers [31], as shown in Fig. 2.1. There are three kinds of CuO_2-layers; consisting of (1) CuO_6 octahedrons, (2) CuO_5 pyramids, and (3) CuO_4 squares as shown in Fig. 2.1. Matrix systems are distinguished by metallic atoms which construct blocking layers or by atoms sandwiched by CuO_2-layers.

In every case, we have a two-dimensional sheet consisting of CuO_2 as a unit. Thus we distinguish oxygen atoms which surround a Cu atom in two ways; in-plane O and apical O. In Fig. 2.2, we show crystal structures of $La_{2-x}Sr_xCuO_4$ (LSCO), $Nd_{2-x-y}Sr_xCe_yCuO_4$ (NSCCO) [32], and $Nd_{2-x}Ce_xCuO_4$ (NCCO) [33] as having the most fundamental structures of cuprates. They all have one CuO_2-layres in each unit cell: CuO_2-layers in LSCO consist of CuO_6 octahedrons, in NSCCO CuO_2-layers consist of CuO_5 pyramids, and CuO_2-layers in NCCO consist of CuO_4 square. These basic structures are called T-phase, T^*-phase, and T'-phase, respectively [33]. It is noted that the first two phases only allow hole-doping while materials with T'-phase allow electron-doping only. We also stress here that all cuprates that have higher T_c than 30 K consist of at least one CuO_6-based or CuO_5-based CuO_2-layer per periodicity.

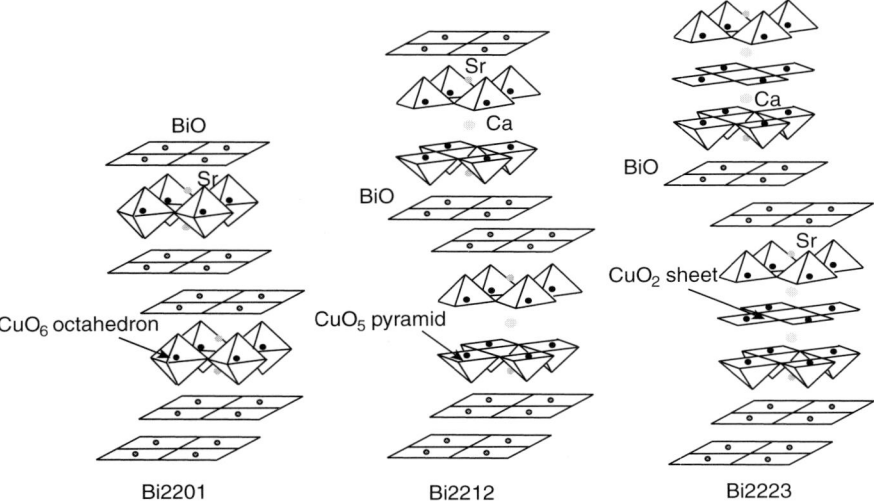

Fig. 2.1. Example of a cuprate family. Here, crystal structures of $Bi_2Sr_2Ca_{n-1}Cu_nO_{4+2n+\delta}$ with $n = 1, 2, 3$ are shown. Oxygen atoms are not drawn for simplicity

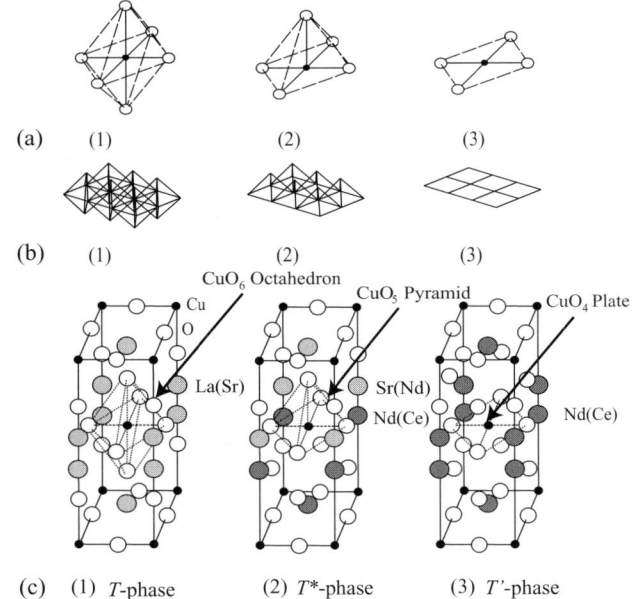

Fig. 2.2. Schematic picture for basic elements of CuO_2-layer in HTSC. (**a**) CuO_6 octahedron, CuO_5 pyramid and CuO_4 square. (**b**) Schematic picture of CuO_2-layers consisting of above mentioned elements. (**c**) Basic crystal structures of HTSC

As for "multi-layered" materials such as $Bi_2Sr_2Ca_{n-1}Cu_nO_{4+2n+\delta}$ (abbreviated as $Bi22(n-1)n$) shown in Fig. 2.1, we observe that there are layers of CuO_5-pyramids face to face separated by metal oxygen (MO)-layers for bi-layer materials like Bi2212. As for Bi2223, we have two layers of CuO_5 pyramids and one CuO_2 layer in a unit cell. There are various kinds of cuprate families distinguished by constituent of metal atoms in the position of blocking layers. Such cuprate families are denoted by those metal atoms, namely LaSrCuO (=LSCO), YBaCuO (=YBCO), BiSrCaCuO (=BSCCO), etc. Among the same families of cuprates, the maximum superconducting transition temperature T_c increases with the number of CuO_2-layers in the unit cell up to $n = 3$. For example, in BSCCO system shown in Fig. 2.1, single-layered Bi2201 has the maximum T_c about 40 K while three-layered Bi2223 has maximum T_c up to 120 K. Some materials allow more than three layers but T_c is found to lower by increasing the number of CuO_2-layers.

As for the material dependence of the maximum T_c, it varies from 40 K for LSCO to 80 K for HgCaCuO for single-layered systems. In addition, the c-axis resistivity ρ_c experiments [34] show that the strength of two-dimensionality is not directly related to the maximum T_c of each materials.

2.1.2 Ability of Changing Carrier Concentration

As we mentioned above, we distinguish the matrix materials by the elements in blocking layers and elements which are sandwiched between CuO_2-layers. Typical matrix systems are given as follows: La_2CuO_4, $YBa_2Cu_3O_{6+x}$, $Bi_2Sr_2Ca_{n-1}Cu_nO_{4+2n}$, $Tl_2Ba_2Ca_{n-1}Cu_nO_{4+2n}$ (TBCCO), and $HgCa_nCu_n$ $O_{1+2n+\delta}$ (with HgCaCuO, no stoichiometric structures are obtained), etc. In any materials, hole carriers are mainly confined to CuO_2-layers, and the insulating states correspond to the states where all Cu atoms in CuO_2-layers are Cu^{2+} ions. A Cu^{2+} ion takes $4s^03d^9$ configuration with the total spin of $S = 1/2$, and this situation is reflected in the experimental findings that all the stoichiometric matrix materials with no hole carrier have three-dimensional antiferromagnetic (3DAF) order whose magnetic transition temperatures T_N are around 240 K~300 K [35, 36, 37].

All cuprate families allow non-stoichiometry. That is, we can change carrier hole concentration continuously by various methods without changing crystalline structure. As for the La_2CuO_4 system, substituting La atoms with Sr atoms from the stoichiometric La_2CuO_4, we have $La_{2-x}Sr_xCuO_4$.

2.2 Experimental Results of Cuprates

In this section we briefly discuss physical properties of superconducting cuprates obtained from experiments. As we have noticed, a tremendous amount of experiments have been performed. Here we concentrate on describing just a few of them which are closely related to the topics of this book.

2.2.1 The Phase Diagram of Cuprates

As we have just mentioned, we can vary the hole concentrations of superconducting cuprates by the substitution of metallic atoms, by adding excess oxygen atoms or by reducing oxygen atoms. The range of the hole concentration, which can be realized by the above-mentioned methods, varies with matrix materials, but it is widely believed that all cuprates have the same phase diagram schematically shown in Fig. 2.3.

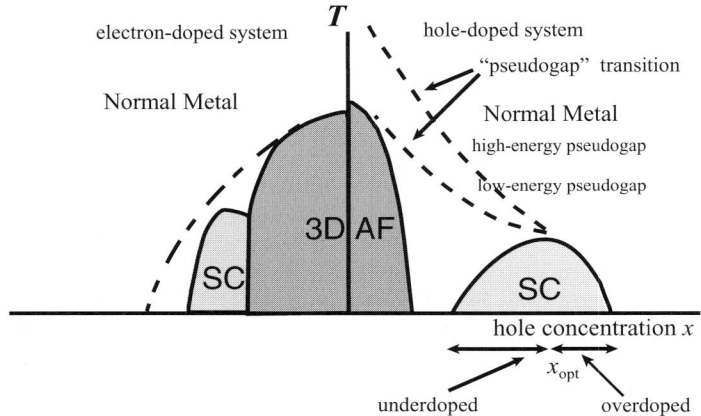

Fig. 2.3. Phase diagram of cuprates. The *dash lines* in the electron-doped system represent the electron-doping concentration dependence of the Néel temperature T_N when oxygen is not reduced

Let us first discuss the phase diagram of hole-doped materials. When the hole concentration x is very small, there exists a static three-dimensional (3D) antiferromagnetic (AF) Néel order. Upon doping, this AF-structure is destroyed rapidly and vanishes at about $x \sim 0.05$ for LSCO, and simultaneously superconductivity (SC) appears. By further doping, the superconducting transition temperature T_c rises but it reaches a maximum value at the concentration $x_{opt} \sim 0.15$ for LSCO [35]. Then upon further doping, T_c decreases and finally it reaches 0 K. For other materials, x_{opt} takes a larger value, but if we estimate the number of holes per CuO_2-layer p by experiments and regard T_c as a function of p, it is known that the optimum p_c where T_c takes the maximum value does not vary so much by materials [38]. As for the maximum value of T_c, it varies depending on matrix materials;

from about 40 K of the first-discovered LSCO to 164 K of the HgCaCuO at 31 GPa. We call the hole-concentration region where T_c goes upward by the increase of hole-concentration "underdoped region", and the region where T_c goes downward, "overdoped" region. The hole-concentration at which T_c reaches a maximum value is called "optimum doping".

The characteristic feature of x-dependence of T_c, $T_c(x)$, is that it is bell-shaped as schematically shown in Fig. 2.3. If the presence of the 3D AF-order at low hole-concentration has a decisive effect on suppressing the occurrence of superconductivity, the $T_c(x)$ curve is expected to rise steeply immediately as the AF-order vanishes. But this is not the case for hole-doped cuprates, because $T_c(x)$ increases gradually with increasing x. This means that the AF-correlation has some role in superconducting properties.

In connection with the x-dependence of T_c, a phase diagram shown in Fig. 2.3 has been suggested for hole-doped cuprates by various experimental results such as NMR [39, 40, 41, 42], ARPES [43, 44, 45], tunneling [46, 47], electronic transports and magnetism [48, 49], electronic specific heat [50, 51], neutron scattering [52, 53], optical properties [54, 55], etc. According to this phase diagram, there are two kinds of "transition-lines" in addition to $T_c(x)$, which are often called "low-energy pseudogap" and "high-energy pseudogap". We will discuss their origins in Chap. 12.

As for electron-doped materials such as $Nd_{2-x}Ce_xCuO_4$, 3D AF-order remains for larger doping concentration as shown schematically in Fig. 2.3. If electron-doped materials are not reduced by oxygen, it is known that no superconductivity occurs at all. By the small reduction of oxygen, 3D AF-order disappear more quickly by electron doping and superconductivity appears immediately after 3D AF-order vanishes. Then it takes the maximum T_c of 23 K at around electron concentration $x \sim 0.18$ for $Nd_{2-x}Ce_xCuO_{4-\delta}$ and T_c becomes zero at around $x \sim 0.22$.

2.2.2 The Symmetry of the Gap

The high resolution of angle resolved photo-emission spectroscopy (ARPES) achieved in the last decade enables us to investigate the $\boldsymbol{k}$-dependence of the gap function of superconducting cuprates. Many experiments suggested that the amplitude of the gap function vanishes along the lines $k_x = \pm k_y$ [44, 56, 57]. Existence of "nodes" on the gap function has been strongly suggested from various other experiments, too. For example, NMR relaxation rate experiments showed that there are no Hebel–Slichter peaks on temperature dependence of relaxation rates $(T_1 T)^{-1}$ and $(T_1 T)^{-1}$ decreases with temperature T in powers of T [58, 59]. In the case of ordinary superconductors, it is known that $(T_1 T)^{-1}$ has a peak just below T_c and then it decreases with temperature T decreasing as $\exp(-\Delta/T)$, reflecting the non-vanishing s-wave gap symmetry (see, for example, [60]). Strong evidence for d-wave symmetry is also obtained from experiments such as penetration depth measurement [61, 62, 63], specific heat [64, 65, 66], Raman scattering [67, 68],

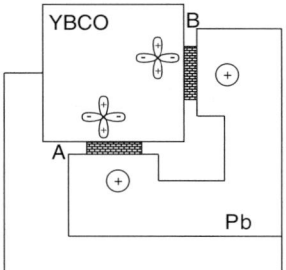

Fig. 2.4. Schematic configuration of π-junction experiment. If HTSC has d-symmetry, cancellation of phase factor causes zero current for zero flux modulo quantum flux while maximum tunneling current is achieved for half quantum flux modulo quantum flux

tunneling [69, 70], etc. These experimental results are naturally understood if we adopt a picture of $d_{x^2-y^2}$-symmetry for the symmetry of the superconducting gap. However, there is one more possibility for the anisotropy of the gap function; the strongly anisotropic s-wave symmetry. We cannot determine either of the symmetry, $d_{x^2-y^2}$-symmetry or the anisotropic s-symmetry (extended s-symmetry), is true for HTSC from ARPES experiments alone.

Distinguishing the $d_{x^2-y^2}$-symmetry from the anisotropic s-symmetry is possible by experimental methods which can detect the phase of the gap function directly. One such experimental method, which uses Josephson tunneling between cuprates and ordinary s-wave superconductors was proposed by Sigrist and Rice [71]. Since it detects the sign change, i.e., the change of phase by π, it is called "π-junction experiment". Schematic configuration of the π-junction experiment is shown in Fig. 2.4. Similar kinds of experiments were actually performed for hole-doped YBCO by Wollam et al. [72] and by Tsuei et al. [73]. Their results support d-wave symmetry and now the same kind of experiments have been done for various kinds of hole-doped cuprates [74, 75]. As for electron-doped materials, early experimental results suggested s-symmetry but recently π-junction experiments were also performed for electron-doped materials [76] and from these experimental results, it seems that cuprates always have $d_{x^2-y^2}$-symmetry, although several arguments have been made for the appearance of other symmetries. (see [77] for example)

So far we have described experimental results which are directly related to the contents of the present book. Experimental results which are specifically related to the K–S model will be discussed in respective chapters.

3 Brief Review of Models of High-Temperature Superconducting Cuprates

3.1 Introduction

In the present chapter we give a brief review of theories of high temperature superconductivity (HTSC). A considerable number of theories have been proposed since the discovery of HTSC, but here we review just a few of them which have a different nature. Roughly speaking, we can classify them into two models: Theories that essentially rely on the Fermi liquid picture and those which presume a much more exotic picture as the basic electronic structure of cuprates. Many theories adopt a view that strong electron-correlation plays an important roll in determining the electronic structure of cuprates, and since there is no standard method to treat strongly correlated electronic systems, theories differ even in the understanding of the normal state of HTSC. In the following we mainly discuss theories with strong electron-correlation.

Before explaining theories of HTSC, let us first look at some basic features of cuprates. As we mentioned, all cuprates have quasi-two dimensional structures with layers consisting of Cu and O, and there are three basic structures for the CuO_2-layer: one consisting of CuO_6 octahedrons, one consisting of CuO_5 pyramids, and one consisting of CuO_4 planes, as shown in Fig. 3.1. In every case, we have a two-dimensional sheet consisting of CuO_2 as a unit. Thus we distinguish oxygen atoms which surround a Cu atom in two ways; in-plane O and apical O.

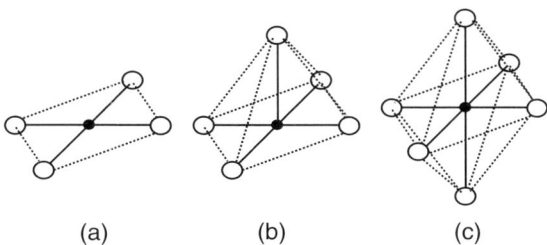

Fig. 3.1. Schematic picture for basic elements of CuO_2-layer in cuprates. (**a**) CuO_4 plane, (**b**) CuO_5 pyramid, and (**c**) CuO_6 octahedron

It is known that in some cuprate materials, such as the first-discovered La-system, undoped materials show no superconductivity and they are even insulators with static antiferromagnetic order. Upon doping, they become metallic and show superconductivity. From a naive point of view, since they have odd numbers of electron per unit cell, they should be half-filled metals. This discrepancy between the "naive" theoretical consideration and experimental results can be easily understood if we take account of the effect of the strong Coulomb repulsion. As shown in Fig. 3.2(a), electrons at each Cu-site cannot be itinerant due to the strong on-site Coulomb repulsion U and also due to the energy difference Δ of one-electron states between the Cu $3d_{x^2-y^2}$ state and O $2p_\sigma$ state. This physical picture also enables us to explain why such insulating materials show anti-ferromagnetism. Virtual processes shown in Fig. 3.2(b) gives rise to an anti-ferromagnetic (AF) superexchange interaction J between neighbouring localized electrons on Cu-sites via intervening O^{2-} ions.

Upon doping, the AF-transition temperature T_N falls quickly and systems become metallic. And by further doping, superconductivity appears and the transition temperature T_c becomes higher and then reaches a maximum value. By further doping, T_c decreases and drops to $0\,\mathrm{K}$. These features are already shown in Fig. 2.3 in Chap. 2.

Now let us introduce various theories proposed for HTSC.

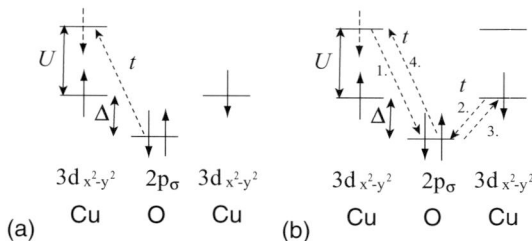

Fig. 3.2. Schematic diagram of the electronic state of the insulating phase in copper oxides. U denotes the on-site Coulomb repulsion (the Hubbard U), t the transfer integral between Cu $3d_{x^2-y^2}$ and O $2p\sigma$ orbitals, and Δ the difference of one-electron energy between Cu $3d_{x^2-y^2}$ and O $2p\sigma$ orbitals. (a) Because of the large value of U, electrons are localized on each Cu site. (b) Schematic picture showing the origin of superexchange interaction J between two neighbouring Cu-site spins. Successive virtual transfer of localized electrons shown in the perturbation processes (4 to 1) apparently favors the antiferromagnetic configuration of the two electrons

3.2 Brief Review of Theories for HTSC

3.2.1 Jahn–Teller Polarons and Bipolarons

A Cu^{2+} ion in a crystal placed in surroundings of octahedral symmetry has a ground state which is orbitally doubly-degenerate. As a result, it is subject to the Jahn–Teller distortion according to the Jahn–Teller theorem [78]. As described in a previous chapter, the crystal structure of undoped La_2CuO_4 is tetragonal at high temperatures. This lower symmetry of La_2CuO_4 has been considered to be due to the Jahn–Teller distortion. Then, when hole-carriers are introduced into La_2CuO_4 by substituting Sr^{2+} ions for La^{3+} ions, apical oxygen in a CuO_6 octahedron in La_2CuO_4 tends to approach Cu ions by the "anti-Jahn–Teller effect" so that a CuO_6 elongated octahedron in La_2CuO_4 deforms in such a way that the Cu-apical O distance is contracted in $La_{2-x}Sr_xCuO_4$ (abbreviated as LSCO hereafter), as we described in a previous section. We call the former and the latter "elongated" and "deformed" octahedrons, respectively. When these hole-carriers interact with distorted CuO_6 octahedrons via electron-lattice interactions, they become polarons. These polarons are called "Jahn–Teller (JT) polarons" [79, 80, 81, 82, 83, 84]. When the transfer integral of a hole-carrier t is much smaller than the Jahn–Teller stabilization energy E_{JT}, a severely confined self-trapped state around an octahedron site is produced. Such a state is called a "small polaron". On the other hand, when t is much larger than E_{JT}, a hole-carrier can distort CuO_6 octahedrons over many lattice distances. Such a hole-carrier is called a "large polaron" [84, 85, 86].

When an electron–phonon coupling constant is large, Alexandrov and Mott [87] argued a possible appearance of a bipolaronic charged Bose liquid. They argued that two JT-polarons form a bounded pair in the real space which they call a "bipolaron", and the superconductivity of cuprates comes from the Bose condensation of these bipolarons in cuprates. From the standpoint that superconducting cuprates are heavily-doped ionic insulators with local distortions, the formations of various types of bipolaron, such as small bipolarons, large bipolarons, JT-bipolarons, etc. have been suggested as a mechanism of high temperature superconductivity [20, 80, 81, 84, 86, 87, 88]. In fact, Bednorz and Müller [3] made a discovery of superconducting cuprates by considering that materials with a strong JT effect may reveal superconductivity with high T_c. So far we have described a qualitative aspect of the bipolaron mechanism. From a quantitative perspective, there are various problems to be clarified. In particular, recent Angle-Resolved-Photoemission (ARPES) experiments by Yoshida and his coworkers [23, 24, 25] clearly showed the existence of "Fermi arcs" for LSCO, which may be considered as the arc section of a Fermi surface. These experimental results are not compatible with bipolaron models in which hole-carriers form bosonic particles. Further, a hopping process from deformed octahedrons to elongated octahedrons sometimes makes polarons localized, or makes the mass of a polaron heavier. As a result a superconducting transition temperature T_c may not be high enough to produce high temperature superconductors. In this book, we will investigate these problems quantitatively.

3.2.2 The Resonating Valence Bond (RVB) State and Quasi-particle Excitations

Assuming that the hole-carriers itinerate in the CuO_2 planes, the RVB state [89] is represented by the single-band Hubbard Hamiltonian,

$$H = \sum_{nm\sigma} t\{c^{\dagger}_{n\sigma} c_{m\sigma} + h.c.\} + \sum U n_{n\uparrow} n_{n\downarrow} , \quad (3.1)$$

where $c_{n\sigma}$ ($c^{(\dagger)}_{n\sigma}$) denotes the annihilation (creation) operator of an electron with spin σ on the Cu $3d_{x^2-y^2}$ (hereafter $d_{x^2-y^2}$) at the $\boldsymbol{n}$th site of a CuO_2-layer, $n_{n\sigma}$ the number operator with spin σ at the $\boldsymbol{n}$th site, while t denotes the transfer between neighbouring Cu $d_{x^2-y^2}$ sites, U the on-site Coulomb repulsion, i.e., the Hubbard interaction between electrons on the same copper site with different spins. As seen from the model Hamiltonian it only considers the highest occupied level of CuO_2-layer, the Cu $d_{x^2-y^2}$ level. If we start from the half-filled state, i.e., the system with one hole per Cu atom, because of the strong on-site Hubbard interaction U, the system is considered to be an antiferromagnetic insulating system with the superexchange interaction $J = t^2/U$. Anderson considered that the ground state of the system is well described by the so-called resonating valence bond (RVB) state, which was first introduced by Anderson himself for other antiferromagnetic Heisenberg spin systems [90]. The RVB state is expressed by the superposition of many configurations of the local singlet pairs (spin singlet states consist of two neighbouring Cu localized $d_{x^2-y^2}$ spins) as illustrated in Fig. 3.3. Reflecting the low dimensionality of CuO_2-layers, localized spins should have a strong quantum fluctuation effect and Anderson proposed the RVB state as a candidate for the ground state of the single Hubbard model of two-dimensional square lattice system at the half-filling. We readily see that the RVB state has no long-range antiferromagnetic order from Fig. 3.3.

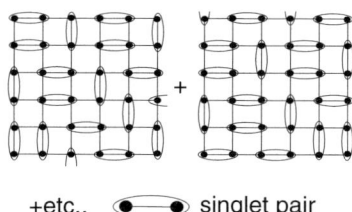

+etc., ●—● singlet pair

Fig. 3.3. Schematic picture for the RVB ground state. Localized Cu spins form a local singlet pair and the RVB ground state is expressed as the superposition of all configurations of states like those illustrated in the figure

3.2 Brief Review of Theories for HTSC

Using the second quantization, the RVB state is written as follows.

$$\Psi_{\text{RVB}} \propto \mathcal{P}_N \left(\sum_{n,m} \delta_{\pm u_i, n-m} c_{n\uparrow}^\dagger c_{m\downarrow}^\dagger \right)^N \Psi_0 , \quad (3.2)$$

where $c_{n\sigma}^\dagger$ is the creation operator of an electron with spin σ at the nth site and $\mathcal{P}_N$ is the projection operator to the fixed total number N state with no doubly occupied states on each Cu atom sites. The u_is with $i = 1, 2$ denote the unit vectors of the square lattice system, which connect neighbouring Cu sites, and the summation for n and m is taken all over the lattice sites, $2N$ being the number of Cu atoms in the system. The summation in (3.2) is taken over all neighbouring Cu sites n and m. On the other hand, it is known that the BCS ground state with N'-Cooper pairs can be written as [60]

$$\Psi_{BCS} \propto \mathcal{P}_{N'} \left(\sum_{n,m} g(n-m) c_{n\uparrow}^\dagger c_{m\downarrow}^\dagger \right)^{N'} \Psi_0 , \quad (3.3)$$

where $g(n)$ is a pair function of the BCS state and $\mathcal{P}_{N'}$ is the projection operator to the fixed total number N' state. One readily sees a strong analogy in the form of two wave functions Ψ_{RVB} and Ψ_{BCS}. From this observation, Anderson argued that the ground state of the hole-doped CuO_2-layer becomes superconducting with the ground state wave function Ψ_{RVB}, which is illustrated in Fig. 3.4.

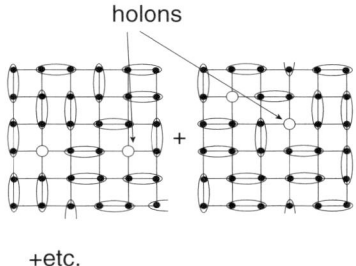

Fig. 3.4. Schematic picture for the RVB ground state *with* a finite hole concentration. In this case, local singlet pairs are formed by all Cu sites except for vacant sites as shown in the figure. In the figure we have two holons

As for the one-particle excitation, Anderson predicted a much more exotic picture. If we introduce a hole in the half-filled RVB state, inevitably there appears an unpaired electron spin and a vacant state with charge $+e$ as seen from Fig. 3.5. From the energetic point of view there is no need for the unpaired spin and the vacant site to be bounded each other. Then Anderson concluded that the unpaired spin and vacant site can itinerate independently

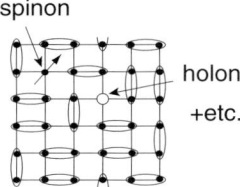

Fig. 3.5. Schematic picture for spinon, holon excitations in the RVB model

with the aid of transfer integral t. As a result we have two kinds of quasi-particle excitations in the RVB state. The neutral spin 1/2 excitation is called "spinon" excitation and the $+e$ charged spin 0 excitation is called "holon" excitation. The ground state of the RVB state for finite hole concentration x (3.3) is then regarded as the Bose condensed state of holon excitations.

The effective Hamiltonian for low-energy phenomena is derived from the single-band Hubbard Hamiltonian (3.1) by an appropriate canonical transformation and it is written as follows.

$$H_{\text{RVB}} = \sum_{nm\sigma} t\{P_n c_{n\sigma}^\dagger c_{m\sigma} P_m + h.c.\} + \sum_{nm} J \boldsymbol{s}_n \cdot \boldsymbol{s}_m , \qquad (3.4)$$

where $P_{\boldsymbol{n}} = 1 - n_{\boldsymbol{n}\uparrow} n_{\boldsymbol{n}\downarrow}$ is the projection operator at the $\boldsymbol{n}$th site to exclude the doubly occupied state and $\boldsymbol{s}_{\boldsymbol{n}}$ denotes the spin operator at the $\boldsymbol{n}$th Cu site if there is just one electron on this site, otherwise the zero operator. The RVB Hamiltonian, which has far less degrees of freedom compared with the single Hubbard Hamiltonian, is still practically unsolvable. The reduction is expressed by $P_{\boldsymbol{n}}$, which has abstract, mathematical form, so that the treatment of this term is a main difficulty in solving the problem. On the other hand, the spinon–holon picture can be understood intuitively. That is, we see some features of the model without solving the problem rigorously. As we see from Fig. 3.6, a holon motion inevitably destroys local singlet coupling of the RVB state. Then it is expected that the effective transfers for

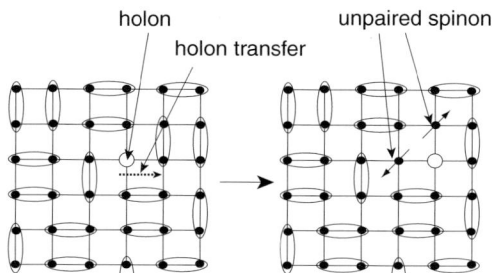

Fig. 3.6. Schematic diagram explaining the frustration effect of a holon motion. If a holon transfers to the neighbouring Cu site, it inevitably destroys the local singlet pair formed in that site, and as a result two unpaired spins appear

holon-excitations are much reduced from the bare t because of this frustration effect. Since the superconducting transition in the RVB state can be treated as the Bose condensation of holons, the reduction of the transfer of a holon, i.e., enhancement of the effective mass of a holon, suppresses the superconducting transition temperature T_c.

3.2.3 The d–p Model

From optical [91, 92] and X-ray absorption (XAS) [93, 94] experiments, it is revealed that O $2p$ orbitals play significant roles in the electronic state of hole-doped systems. Taking this fact into account, Emery [95] proposed a model Hamiltonian which includes O $2p_\sigma$ orbitals in a CuO$_2$-layer. Because this model considers not only the Cu $d_{x^2-y^2}$ orbitals but also in-plane O $2p_\sigma$ orbitals, it is called the d–p model. A schematic picture of the d–p model is given in Fig. 3.7. The figure is written from the hole picture so that the "one electron" level of O $2p_\sigma$ orbital is higher than that of Cu $d_{x^2-y^2}$ and all the O $2p_\sigma$ levels are empty at the half filling state. When we introduce one hole in this system, it is accommodated in an O $2p_\sigma$ orbital or an upper Hubbard level of Cu $d_{x^2-y^2}$ as shown in Fig. 3.7(a). Then a hole becomes itinerant by the mixing of these two levels, which form a band with typical energy level between the O $2p_\sigma$ level and the upper Hubbard Cu $d_{x^2-y^2}$ level. Emery and others who support the d–p model assigned this band to the so-called "mid-gap state" found in optical and X-ray absorption experiments of cuprates [91, 93]. The important point is that this "mid-gap state" energy

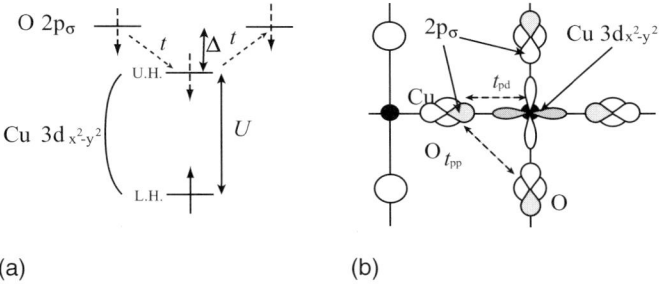

Fig. 3.7. Schematic diagram describing the d–p model in the hole picture. (**a**) Schematic diagram for the Hubbard interaction U, difference of one-electron level Δ between O $2p_\sigma$ and Cu $d_{x^2-y^2}$, and the transfer integral t_{dp} between O $2p_\sigma$ and Cu $d_{x^2-y^2}$ in the d–p model. *Solid arrows* represent localized spins while a *dashed arrow* represents a doped hole. UH denotes the upper Hubbard level and LH the lower Hubbard level of Cu $d_{x^2-y^2}$. Here, the hole picture is taken so that empty orbitals of O $2p_\sigma$ levels correspond to the fully occupied state in the ordinary picture. (**b**) Schematic picture of O $2p_\sigma$ orbitals and Cu $d_{x^2-y^2}$ orbital, and the transfer t_{dp} between O $2p_\sigma$ and Cu $d_{x^2-y^2}$, the transfer t_{pp} between O $2p_\sigma$s are shown

is much smaller than the estimated Hubbard U of Cu $d_{x^2-y^2}$, $U \sim 10\text{eV}$, and it is close to the estimated energy difference of Cu $d_{x^2-y^2}$ and O $2p_\sigma$, a few eV. This suggests that doped holes have strong O $2p_\sigma$ character.

The d–p model itself does not concern the origin of the superconductivity of cuprates. It explains the basic electronic structure of copper oxides. As for the mechanisms of HTSC, there are two different approaches: One from the standpoint of strongly correlated interaction and the other from, essentially, the viewpoint of weakly correlated Fermi liquid. The former approaches derive the effective Hamiltonians from the d–p model Hamiltonian, provided that the ground state of the half-filled system of the d–p model is an insulating Heisenberg antiferromagnet. The latter approaches are essentially based on the conventional Fermi liquid picture. They are based on the perturbation expansion, i.e., treating the Hubbard term as the perturbed Hamiltonian. These approaches are described in the subsequent subsections.

3.2.4 The t–J Model

As we have seen in the preceding subsection, if we introduce a hole in the ground state of the d–p model at the half filling, we have a state which is written as the superposition of O $2p_\sigma$ and upper Hubbard Cu $d_{x^2-y^2}$. Zhang and Rice [96] argued that this state is written as the superposition of local singlet states; namely, a dopant hole around a Cu site form a singlet state with a localized Cu hole spin, as illustrated in Fig. 3.8, and this singlet state (Zhang–Rice singlet) itinerates in the CuO$_2$-plane. They also showed that the effective Hamiltonian of the model can be derived from the d–p model by an appropriate approximation. The resultant effective Hamiltonian has the following form:

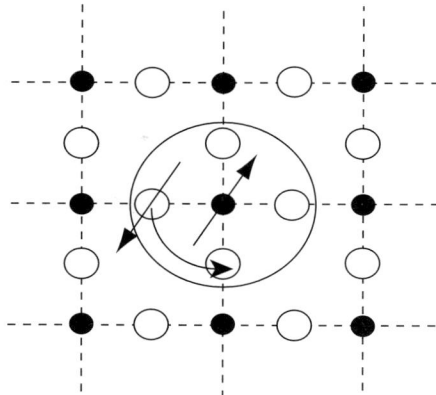

Fig. 3.8. Schematic picture of Zhang–Rice singlet. A dopant hole in O $2p_\sigma$ orbitals surrounding a Cu site forms a spin singlet with the localized spin on the Cu site

$$H_{t-J} = H_t + H_J$$
$$= \sum_{n\,m} t\{c_{n\,\sigma}^\dagger c_{n\,\sigma} + h.c.\} + \sum_{n\,m} J\boldsymbol{S}_n \cdot \boldsymbol{S}_m \qquad (3.5)$$

where $S_{n\,x} + iS_{n\,y} = c_{n\,\uparrow}^\dagger c_{n\,\downarrow}$, $S_{n\,z} = \{n_{n\,\uparrow} - n_{n\,\downarrow}\}/2$.

Since this Hamiltonian consists of the transfer part H_t and antiferromagnetic interaction part H_J, it is called the t–J Hamiltonian, and a system described by H_{t-J} is called "t–J" model. The itinerancy of the Zhang–Rice singlet state as a quasi-particle state can be described by the "t–J Hamiltonian", H_{t-J} in (3.5).

Since the t–J Hamiltonian is derived from the effective Hubbard Hamiltonian (3.5), the exotic picture of quasi-particle excitation of the RVB state is inherited to the t–J model. Then there are two kinds of condensation transition. Namely, BCS like spinon pairing condensation and the Bose condensation of the holon system. From the mean-field calculation for the t–J model, the spinon pairing transition temperature T_S is found to decrease with increasing hole concentration. On the other hand, the Bose condensation temperature T_B increases as the hole concentration increases [97]. The superconducting transition is considered to be realized in the temperature region where both Fermion (spinon) condensation and Boson (holon) condensation occur. As a result we obtain the "bell-shaped" character for the $T_c(x)$-curve from the t–J model.

Moreover, the t–J model can explain the pseudogap behaviour of observed experiments by assigning a pseudogap transition to the Fermion condensation line. The phase diagram shown in Fig. 3.9 looks consistent with experimental results. These results are mainly obtained by the mean field approximation to the spinon and holon excitations with the slave-boson or the slave-fermion

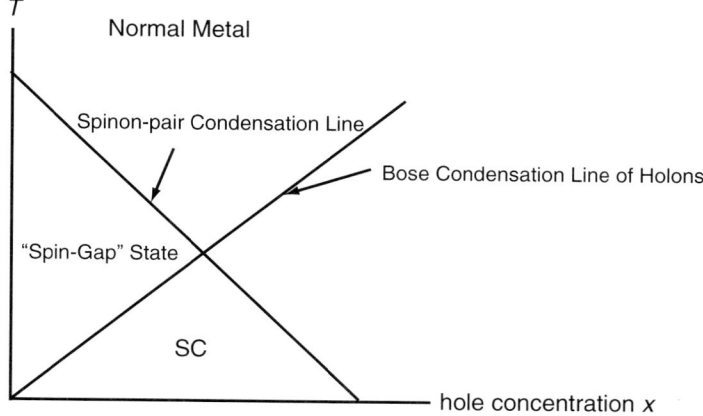

Fig. 3.9. Phase diagram of the t–J model obtained from the mean field approximation [97]

method. However, there are critical opinions for using the simple mean-field approximation to the exotic spinon–holon picture. In the mean-field approximation for the slave-boson or slave-fermion method, for example, the sum of the average numbers of spinons and holons must be one. From this, the average number of spinons becomes large in the underdoped region of holons which contribute to superconductivity. According to the RVB theory and thus the t–J model, the spinons contribute to the formation of large Fermi surfaces. Thus the appearance of a large Fermi surface is expected for the underdoped region. However, recent ARPES experiments by Yoshida et al. [23, 24, 25] clearly shows that this is not the case. In this context, the justification of the t–J model by the mean-field approximation is questionable unless one can obtain a direct experimental evidence for the existence of spinon and holon excitation.

3.2.5 Spin Fluctuation Models

Starting from the single-band Hubbard model [98, 99, 100], the d–p model [101], or an electron dispersion obtained from the "LDA-band calculations plus Hubbard U" (the LDA+U band calculations) by treating in a semi-empirical formalism [102], another model has been introduced as regards the mechanism of high-T_c.

Near the half-filled level, strong Hubbard U repulsion between electrons causes the enhanced $\boldsymbol{q},\omega$-dependent spin susceptibility $\chi(\boldsymbol{q},\omega)$. It takes a large value around $\boldsymbol{Q} = (\pi/a, \pi/a)$ where a denotes the lattice constant for the CuO$_2$ plane. This fact reflects the nearly antiferromagnetic nature of the system. This large spin fluctuation causes an "overall" attractive interaction between quasi-particles with different spins, thus leading to the superconducting transition. Theories depending on such mechanisms are called "spin fluctuation" models. Here we use the word "overall" in the above sentence because the interaction caused by the antiferromagnetic spin fluctuation has both repulsive and attractive components and it appears that in the $\boldsymbol{k}$-space it is always positive, i.e., *repulsive*. From the perturbation theory, the dressed electron–electron interaction $\mathcal{V}_{eff}$ is written in terms of U and $\chi(\boldsymbol{q},\omega)$, and its $\boldsymbol{q}$-dependence is similar to that of $\chi(\boldsymbol{q},\omega)$, i.e., it takes a small value around $\boldsymbol{q} = 0$, while it takes a very large value at $\boldsymbol{q} = \boldsymbol{Q} = (\pi/a, \pi/a)$. Together with the shape of the Fermi surface which has a large partial density of states around $(\pi/a, 0)$ and $(0, \pi/a)$, we obtain d$_{x^2-y^2}$-wave superconductivity which is consistent to the experimental results.

There are various models starting from the above mentioned picture. In other words, these theories identify the origin of Cooper pair interaction in HTSC with a large anti-ferromagnetic spin fluctuation of the system, but they differ in their detail; they differ in the approximation methods to solve a similar type of model Hamiltonians. These theories can explain normal state properties of cuprates such as magnetic properties or optical responses. In most of the theories of this category, they rely on strong antiferromagnetic fluctuation effect so that T_c inevitably becomes larger as the hole concentration decreases. This is apparently against the experimental results. One idea to avoid this discrepancy between theories and experimental results is to include terms in the perturbation expansion which we usually do not count in a simple approximation, i.e., in the random phase approximation (RPA). [100, 101]. In any case, some self-consistent conditions for the Geen's function, the proper self-energy part and the irreducible vertex part are derived, and to solve equations for these many-body correlation functions with self-consistent conditions is a main task in this field.

3.2.6 The Kamimura–Suwa Model and Related Two-Component Mechanisms

3.2.6.1 On the Kamimura–Suwa (K–S) Model

Theories so far reviewed start by assuming that the hole-carriers move in a CuO_2 plane, followed by a refinement which makes use of disposal parameters to fit experimental data. This procedure makes it difficult to assess the predictive nature of the model, for example, how cuprates containing a CuO_6 octahedron or a CuO_5 pyramid may give rise to different features of high T_c superconductivity. In order to address this problem, Kamimura and his coworkers have carried out a series of theoretical studies, beginning with the calculations of the electronic structure of LSCO from first principles by Kamimura and Eto [103, 104]. In these first principles calculations, Kamimura and Eto took account of the local distortion of a CuO_6 octahedron when Sr^{2+} ions are substituted for La^{3+} ions in LSCO. That is, when La^{3+} ions are replaced by Sr^{2+} ions, apical oxygen in CuO_6 octahedrons tend to approach Cu ions so as to gain the attractive electrostatic energy. Such a contraction effect of the apical O–Cu distance in a CuO_6 octahedron by doping was predicted theoretically by Shima and his coworkers in 1988 [106] and supported by various experimental groups by the neutron time of flight experiments in 1990s [107, 108, 109]. As a result the octahedrons which were elongated along the c-axis by the Jahn–Teller effect in the undoped La_2CuO_4 shrink by doping the divalent ions for the trivalent ions. We call this local distortion of CuO_6 octahedrons "anti-Jahn–Teller effect". A similar anti-Jahn–Teller effect also occurs in other cuprates in which the hole-doping is caused by the excess of oxygen in blocking layers. Thus CuO_6 octahedrons or CuO_5 pyramids in cuprates are deformed in a form of contraction or elongation along the c-axis.

Kamimura and his coworkers have tried to solve the many-electron lowest state called "multiplet" of a CuO_6 or a CuO_5 cluster in cuprates as accurately as possible by Multi-Configuration Self-Consistent Field Method with Configuration Interaction (MCSCF-CI), by taking account of the anti-Jahn–Teller effect. It has been shown that the anti-JT effect plays an important role in introducing two kinds of multiplets (or orbitals) which are very close in their energies. The results of first-principles cluster calculations mentioned above have led to the Kamimura–Suwa model. Based on the results of MCSCF-CI calculations, Kamimura and Suwa showed that the hole-carriers in the underdoped regime of cuprates form a metallic state, by taking the Zhang–Rice singlet and the Hund's coupling triplet alternately in the presence of the local AF order constructed by the localized spins in a CuO_2 plane. The metallic state in cuprates constructed in the above-mentioned way is nowadays called the "Kamimura–Suwa model" which is abbreviated as the K–S model. We may say that the K–S model was the onset of "two-component theories", whose various modifications are nowadays adopted by a number of theoretical models.

Theoretically the result of the LDA + U band calculation by Anisimov, Ezhov and Rice [110] has supported the K–S model, while experimental evidence for the K–S model has been reported by Chen and coworkers [93] and by Pellegrin and coworkers [94] independently by performing the experiments of polarized X-ray absorption spectra for LSCO, and also by Merz and coworkers [111] by the experiment of site-specific X-ray absorption spectroscopy for $Y_{1-x}Ca_xBa_2Cu_3O_{7-y}$.

In order to solve the K–S model quantitatively, Kamimura and Ushio [112, 113] proposed the mean-field treatment for the exchange interaction between the spins of hole-carriers and of the localized holes in the same CuO_6 octahedron (or CuO_5 pyramid), which is a very important interaction in the K–S model.

By applying the mean-field treatment to the K–S effective Hamiltonian, the above exchange interaction can be expressed as a form of an effective magnetic field acting on the carrier spins. In this way the carrier system and the localized spin systems can be separated. As a result, the electronic structure of a hole-carrier system on the K–S model can be expressed in a form of a single-electron-type band structure in the presence of AF order in the localized hole-spin system, where the single-electron-type band structure includes many-body effects such as the exchange interaction between the spins of a hole-carrier and of a localized hole in the mean-field sense. Based on the mean-field approximation for the localized spin system by Kamimura and Ushio, the many-body-effect including energy band, Fermi surfaces and the density of states can be calculated. Further thermal, transport and optical properties of the underdoped cuprates can be calculated by the above-mentioned energy bands and wavefunctions, and the calculated results are compared with experimental results. Summarizing the description of the K–S model, we can say that the K–S model in the underdoped regime of cuprates has the following two important features; (1) two-component scenario in a metallic state such as the coexistence of Zhang–Rice singlet and Hund's coupling triplet in the presence of local AF order and (2) inhomogeneous superconducting state due to inhomogeneous charge distribution in a metallic state. With respect to the first feature, one may also say the coexistence of two kinds of orbital, the b_{1g} bonding orbital consisting of mainly four oxygen p_σ orbitals in a CuO_2 plane and the a_{1g} antibonding orbital consisting of mainly Cu d_{z^2} orbital from the standpoint of orbital characters. Similar two-component mechanisms have been developed by several theoretical groups by considering both $d_{x^2-y^2}$ and d_{z^2} orbitals [114, 115, 116, 117, 118]. In particular, Bussmann-Holder and her coworkers [114, 116] have developed a two-component theory similar to the K–S model by taking account of the orthorhombic distortions.

3.2.6.2 Is a Superconducting State in Cuprates Homogeneous or Inhomogeneous?

As regards the second feature, a controversial question of whether the superconducting state in cuprates is inhomogeneous or not has been recently raised. Concerning this question, a number of international conferences on inhomogeneity and stripes have been held in various places and a considerable number of papers have been published. With regard to the K–S model, as we shall describe later in this book, a metallic state is certainly inhomogeneous in the sense that a metallic state is bounded by spin-flustration regions on the boundary of an AF spin-correlated region. A superconducting as well as a metallic state are extended by the spin-fluctuation effect in a two-dimensional AF Heisenberg spin system like percolation. Thus this superconducting state is surrounded by insulating regions. In this sense the superconducting state on the K–S model may be said to behave like a dynamical stripe. Since Tranquada and his coworkers [119] reported in 1995 on the dynamical modulation of spin and charge in LSCO, the static and/or dynamic charge and/or spin stripes have been the subject of many theoretical and experimental studies. Since the aim of this chapter is not to review these works, here it suffices to mention references related to work on stripes and inhomogeneity. Here we will mention only the name of conference proceedings which one of the present authors attended, rather than mentioning a huge number of individual papers. Those are the first and third international conferences on stripes and high T_c superconductivity, which were held in Rome in 1996 [120] and 2000 [121], respectively. Further we would also like to mention papers published in Sect. 3.8 entitled "Stripe Phase and Charge Ordering" in the proceedings of the international conference on Materials and Mechanisms of Superconductivity High Temperature Superconductors VI, Part III, which was held in Houston in 2000 [122].

4 Cluster Models for Hole-Doped CuO_6 Octahedron and CuO_5 Pyramid

4.1 Ligand Field Theory for the Electronic Structures of a Single Cu^{2+} Ion in a CuO_6 Octahedron

The crystal structure of La_2CuO_4 is tetragonal at high temperatures and is of a layer-type. In this crystal structure a CuO_2 unit forms a square planar network in each layer (on x–y plane) perpendicular to the c-axis, as seen in Fig. 4.1, where each Cu^{2+} ion is surrounded by six O^{2-} ions nearly octahedrally. The many-electron ground state called "multiplet" for a Cu^{2+} ion ($3d^9$), placed in surroundings that are of octahedral symmetry, is ^2E_g, orbitally doubly-degenerate with the basis functions d_{z^2} and $d_{x^2-y^2}$. Thus it is subject to the Jahn–Teller distortion. As a result this CuO_6 octahedron is stretched along the c-axis, producing two long (2.41 Å) and four short (1.89 Å) Cu–O lengths [123] by Jahn–Teller effect [78].

In a crystalline field with tetragonal symmetry, ^2E_g state is further split into $^2A_{1g}$ and $^2B_{1g}$, where A_{1g} and B_{1g} are the irreducible representation of the D_{4h} group. In La_2CuO_4, a Cu^{2+} ion in a CuO_6 octahedron is mainly subject to a crystalline field with tetragonal symmetry, so that the five-fold degenerate d orbitals of the Cu^{2+} ion with $3d^9$ electron configuration are split into b_{1g}, a_{1g}, b_{2g} and e_g orbitals as shown in Fig. 4.2, where the behaviour of the orbital splitting by the cubic field is also shown. Thus a hole occupies an

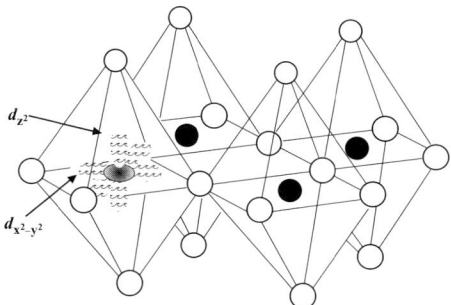

Fig. 4.1. A CuO_2 plane which forms a planar network in each layer. The Cu $d_{x^2-y^2}$ and d_{z^2} orbitals are drawn together

30   4 Cluster Model for Hole-Doped CuO_6 Octahedron and CuO_5 Pyramid

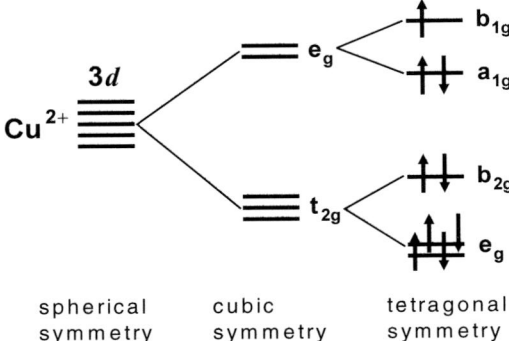

Fig. 4.2. The energy splitting of Cu^{2+} orbitals in spherical, cubic and tetragonal symmetry

antibonding b_{1g} orbital, denoted by b_{1g}^*, which is mainly constructed of the Cu $d_{x^2-y^2}$ orbital with the hybridization of surrounding four O p_σ orbitals in a CuO_2 plane.

4.2 Electronic Structures of a Hole-Doped CuO_6 Octahedron

When dopant holes are introduced in La_2CuO_4, there are two possibilities for orbitals to accommodate a dopant hole in CuO_6. One case is that a dopant hole occupies an antibonding a_{1g}^* orbital consisting of the Cu d_{z^2} orbital and the surrounding six oxygen p_σ orbitals, and its spin becomes parallel by Hund's coupling with localized spin of $S = 1/2$ around a Cu site, which occupies the b_{1g}^* orbital. This spin–triplet multiplet is called "Hund's coupling triplet", denoted by $^3B_{1g}$, as shown in Fig. 4.3(a). The other case is that a dopant hole occupies a bonding b_{1g} orbital consisting of four in-plane oxygen p_σ orbitals with a small Cu $d_{x^2-y^2}$ component, and its spin becomes antiparallel to the localized spin in the b_{1g}^* orbital as shown in Fig. 4.3(b). This multiplet is denoted by $^1A_{1g}$. Since the $^1A_{1g}$ state corresponds to the spin–singlet state proposed by Zhang and Rice, which is also a key constituent state in the t–j model [96], we call the $^1A_{1g}$ multiplet the "Zhang–Rice singlet".

4.3 Electronic Structure of a Hole-Doped CuO_5 Pyramid

Like LSCO, when holes are doped in superconducting $YBCO_7$ with $T_c = 90\,K$ or Bi2212 with $\delta = 0.25$ with $T_c = 80\,K$, there are two possibilities for orbitals to accommodate a dopant hole in a CuO_5 pyramid. One case is that a dopant

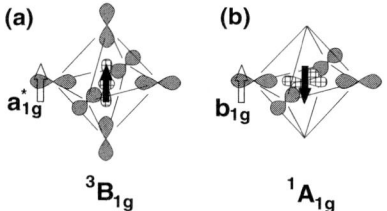

Fig. 4.3. (**a**) Schematic view of $^3B_{1g}$ multiplet called the Hund's coupling triplet in a CuO$_6$ octahedron. A *solid arrow* represents a localized spin while an *open arrow* the spin of a hole carrier which occupies an antibonding a^*_{1g} orbital shown in the figure. (**b**) Schematic view of $^1A_{1g}$ multiplet called the Zhang–Rice singlet, in which a hole occupies a bonding b_{1g} orbital shown in this figure

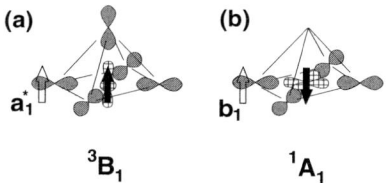

Fig. 4.4. (**a**) Schematic view of $^3B_1$ multiplet, called the Hund's coupling triplet, in a CuO$_5$ pyramid. A *solid arrow* represents a localized spin while an *open arrow* the spin of a hole carrier which occupies an antibonding a^*_1 orbital shown in the figure. (**b**) Schematic view of $^1A_1$ multiplet called the Zhang–Rice singlet, in which a hole occupies a bonding b_1 orbital shown in this figure

hole occupies an antibonding a^*_1 orbital consisting of a Cu d_{z^2} orbital and five surrounding oxygen p_σ orbitals, and its spin becomes parallel to a localized spin of $S = 1/2$ around a Cu site, by Hund's coupling. This multiplet is the "Hund's coupling triplet" denoted by $^3B_1$, as shown in Fig. 4.4(a). The other case is that a dopant hole occupies a bonding b_1 orbital consisting of four in-plane oxygen p_σ orbitals with a small Cu $d_{x^2-y^2}$ component, and its spin becomes anti-parallel to the localized spin as shown in Fig. 4.4(b). This multiplet is the "Zhang–Rice singlet" denoted by $^1A_1$.

4.4 Anti-Jahn–Teller Effect

As we have already described, a regular octahedron in La$_2$CuO$_4$ is elongated along the c-axis by the Jahn–Teller-effect. In order to investigate whether the Jahn–Teller interaction still plays a role in the local distortion of the CuO$_6$ octahedron in the case of hole-doped cuprates, Shima, Shiraishi, Nakayama, Oshiyama and Kamimura calculated in 1987 the optimized Cu–apical O distance in the CuO$_6$ octahedrons by minimizing the total energy of doped La$_{2-x}$Sr$_x$CuO$_4$ as a function of hole-concentration x up to $x = 1.0$. The calculation was performed by the first-principles norm-conserving pseudopotential method within the local density functional formalism (LDF) [106]. The doping effect of Sr is treated by adopting the virtual crystal approximation for disordered $La_{2-x}Sr_xCuO_4$ crystal, in which the virtual atoms consisting of La and Sr are placed with the mixing ratio of $(2-x)$ to x for La to Sr atom on all the La sites. This approximation is expected to be reasonably accurate for such alloy systems of La and Sr ions whose ionic radii are nearly the same. As the lattice constants and the Cu–O(x,y) distance in the CuO$_2$ plane changes little from those in pure La_2CuO_4 even in the case of Sr doped $La_{2-x}Sr_xCuO_4$, all the lattice parameters except the Cu–apical O distance along the z axis are fixed to those values in pure La_2CuO_4; d(Cu–O(x,y)) = 1.89 Å, d(Cu–La) = 4.78 Å, and the distance of c-axis = 13.25 Å.

In Fig. 4.5 the total energies thus minimized are plotted as a function of Cu–O(z) distance for $X = 0.0, 0.1, 0.3$ and 0.5, which correspond to $x = 0.0, 0.2, 0.6$ and 1.0 respectively, in the ordinary formula $La_{2-x}Sr_xCuO_4$. Minimum positions are indicated by arrow for each X. In Fig. 4.6 the optimized Cu–O(z) distances as a function of X are presented. As shown in this figure, the optimized Cu–O(z) distance is 2.41 Å at $X = 0$ and 2.34 Å at $X = 0.1$. When the hole-concentration x increases beyond $X = 0.3$, the Cu–O(z) distance begins to decrease rapidly, and when X is 0.5, at which all Cu ions in LaSrCuO$_4$ becomes Cu^{3+}, the Cu–apical O(z) distance becomes

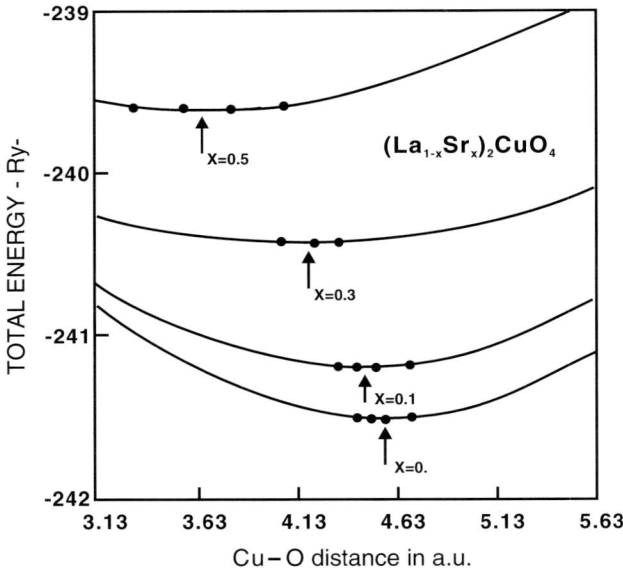

Fig. 4.5. The calculated total energy as a function of the Cu–apical O distance for each value of Sr content x in $La_{2-x}Sr_xCuO_4$ (after Shima et al. [106]), where it should be noted that X in the figure corresponds to $x/2$

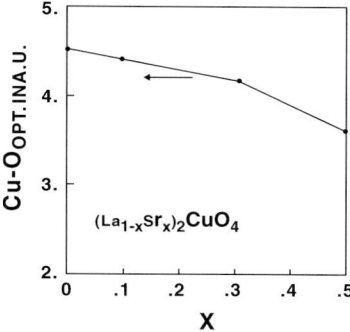

Fig. 4.6. The optimized Cu–apical O distance as a function of Sr content x in $La_{2-x}Sr_xCuO_4$ (after Shima et al. [106]), where it should be noted that X in the figure corresponds to $x/2$

1.89 Å which is the same value as the Cu–O(x,y) distance. The calculated result of 2.41 Å for $X = 0.0$ coincides well with the observed value of the Cu–apical O distance in the undoped La_2CuO_4. Thus the calculated results in Figs. 4.5 and 4.6 clearly indicate that the Cu–apical O distance in the undoped La_2CuO_4 is elongated up to 2.41 Å by the Jahn–Teller interaction. Then, when the hole-concentration x increases, all the Jahn–Teller elongated CuO_6 octahedrons shrink by doping divalent Sr ions for trivalent La ions. The calculated result by Shima et al. showed clearly that the contraction of the Cu–apical O distance from the Jahn–Teller elongated value of 2.41 Å occurs so as to gain the attractive electrostatic energy in the presence of the virtual atoms with character of $(2-x)$ La and x Sr. As a result the total energy gained by the Jahn–Teller effect was reduced by the doping effect in a virtual crystal of $La_{2-x}Sr_xCuO_4$. From this fact we call the contraction effect of the Cu–apical O distance due to doping the "anti-Jahn–Teller effect". In the following chapter we will show by first-principles calculations that the lowest energies of the Hund's coupling triplet and the Zhang–Rice singlet become nearly the same through the anti-Jahn–Teller effect. Thus the two-component system such as the coexistence of the Hund's coupling triplet and the Zhang–Rice singlet becomes essentially important in forming a superconducting state as well as a metallic state in superconducting cuprates.

4.5 Cluster Models and the Local Distortion of a Cluster by Doping Carriers

In the following chapters we will calculate the lowest energies of the Hund's coupling triplet and the Zhang–Rice singlet by taking account of the anti-Jahn–Teller effect. For this purpose we first have to develop a new method of first-principles calculation for a cluster system. Simultaneously, it is necessary to set up a cluster model for first-principles calculations. As such a model for cluster calculations, we adopt a CuO_6 cluster embedded in the LSCO compound, and a CuO_5 cluster embedded in $YBCO_7$ or Bi2212 compounds. We label the oxygen in a CuO_2 plane as O(1), and the apical oxygens as O(2). We use the lattice constants reported in [124] for LSCO and in [125] for $YBCO_7$. The number of electrons is determined so that a formal charge of copper is $+2e$ and that of oxygen is $-2e$ for an undoped case. Then we consider hole-doped systems for LSCO and YBCO by subtracting one electron.

To include the effect of the Madelung potential from the exterior ions outside a cluster under consideration, the point charges are placed at exterior ion sites in a way of $+2e$ for Cu and Ba, $-2e$ for O, and $+3e$ for La, Y and Bi. The number of point charges considered in the first principle calculations is 168 for CuO_6 and 300 for CuO_5. These point charges determine the Madelung potential at Cu, O(1) and O(2) sites within a cluster. Kondo [126] also pointed out this important role of the Madelung potential,

In superconducting cuprates we take account of the effects of the local distortions of a CuO_6 octahedron or CuO_5 pyramid from its elongated form by the anti-Jahn–Teller effect, which plays an important role in determining the lowest state of a CuO_6 octahedron or CuO_5 pyramid. Thus the contraction effect in the distance between the Cu atom and the apical oxygens, which are located above (and below) the Cu atom in the CuO_2 planes, can be taken into account seriously in our theoretical calculations for the present two-component system. So far any model has not seriously considered such distortion effect due to the anti-Jahn–Teller effect, except our calculations [103, 104]. Recently a number of experimental results indicate that the distance between the apical O atom and the Cu atom is reduced when holes are doped into superconducting cuprates such as LSCO [108, 123], YBCO [107, 109] and Bi2212 with $\delta = 0.25$ [5, 127], supporting the theoretical prediction by Shima et al.

In the case of a CuO_6 cluster in LSCO, Kamimura and Eto [104] varied the Cu–apical O(2) distance c, according to the experimental results by Boyce et al. [123] and to the theoretical result by Shima et al. [106]. The distance c is taken as 2.41 Å, 2.35 Å, 2.30 Å and 2.24 Å, depending on the Sr concentration where 2.41 Å and 2.30 Å correspond to the value of c in the cases of $x = 0.0$ (undoped) and of $x = 0.2$, respectively, in the $La_{2-x}Sr_xCuO_4$ formula. In the case of a CuO_5 cluster, on the other hand, the Cu(2)-apical O(2) distance is taken as 2.47 Å for insulating $YBCO_6$ and 2.29 Å for superconducting $YBCO_7$ with $T_c = 90$ K following the experimental results by neutron [107] and X-ray [109] diffraction measurements, where Cu(2) represents Cu ions in a CuO_2 plane while Cu(1) represents Cu(1) ions in a Cu–O chain. In the case of superconducting Bi2212 with $\delta = 0.25$ ($T_c = 80$ K), the distance between Cu

4.5 Cluster Models and the Local Distortion of a Cluster by Doping Carriers

and apical O in the CuO_5 pyramid is 2.15 Å [127], and it is surprisingly short. On the other hand, the distance between Cu and O(1) in a CuO_2 plane is 1.91 Å, which is nearly equal to that of La_2SrCuO_4.

Since we have set up a cluster model for LSCO, YBCO and Bi2212, we are going to proceed to first-principles calculations to calculate the lowest-state energy of each cluster system in the following several chapters.

5 MCSCF-CI Method: Its Application to a CuO$_6$ Octahedron Embedded in LSCO

5.1 Description of the Method

In order to calculate the lowest energy of a CuO$_6$ cluster or a CuO$_5$ pyramid embedded in cuprates based on the cluster models for LSCO, YBCO and Bi2212 set up in Chap. 4, we adopt a method of Multi-Configuration Self-Consistent Field with Configuration Interaction (MCSCF-CI), which was developed by Eto and Kamimura in 1987 [103, 104].

The MCSCF-CI method [128, 129, 130] is the most suitable variational method to calculate the ground state of a strongly correlated cluster system. Kamimura and Eto applied this method to LSCO to calculate the electronic structure of a CuO$_6$ octahedron embedded in LSCO for the first time. Then, by applying this method to Cu$_2$O$_{11}$ dimer in undoped La$_2$CuO$_4$, Eto and Kamimura showed [103, 104] that the holes are localized around Cu sites and these localized holes form a spin–singlet state corresponding to the Heitler–London states in a H$_2$ molecule. This result is consistent with the experimental results of Mott–Hubbard insulator for La$_2$CuO$_4$.

Later Kamimura and Sano [131] and Tobita and Kamimura [132] applied the MCSCF-CI method to YBCO and Bi2212, respectively, and calculated the electronic structure of a CuO$_5$ pyramid embedded in YBCO$_7$ and Bi2212 with $\delta = 0.25$. In this section we give a brief review of how to use this method for the calculations of the lowest state energies of the $^1A_{1g}$ (or $^1A_1$) multiplet in the case of a CuO$_6$ octahedron (or CuO$_5$ pyramid) and $^3B_{1g}$ (or $^3B_1$) multiplet. A variational trial function for the Zhang–Rice singlet $^1A_{1g}$ (or $^1A_1$) in the MCSCF-CI method is taken as,

$$\Phi_S = C_0 |\psi_1\alpha\psi_1\beta\psi_2\alpha\psi_2\beta\cdots\psi_n\alpha\psi_n\beta| \\ + \sum_i \sum_a C_{ii}^{aa} |\cdots\psi_{i-1}\alpha\psi_{i-1}\beta\psi_{i+1}\alpha\psi_{i+1}\beta\cdots\psi_a\alpha\psi_a\beta|, \quad (5.1)$$

while that for the Hund's coupling triplet $^3B_{1g}$ (or $^3B_1$) is chosen as,

$$\Phi_T = C_0 |\psi_1\alpha\psi_1\beta\cdots\psi_{n-1}\alpha\psi_{n-1}\beta\psi_p\alpha\psi_q\alpha| \\ + \sum_i \sum_a C_{ii}^{aa} |\cdots\psi_{i-1}\alpha\psi_{i-1}\beta\psi_{i+1}\alpha\psi_{i+1}\beta\cdots\psi_a\alpha\psi_a\beta\psi_p\alpha\psi_q\alpha|, \quad (5.2)$$

where $2n$ is the number of the electrons in the clusters, and $|\cdots\cdots|$ represents a Slater determinant. For example, $2n = 86$ for a CuO_6 octahedron cluster, and $2n = 76$ for a CuO_5 pyramid cluster. Orbitals ψ_p and ψ_q in (5.2) are always singly occupied. In (5.1) and (5.2) all the two-electron configurations are taken into account in the summation over i and a so that the electron-correlation effect is effectively included in this method. By varying ψ_i's and coefficients C_0 and C_{ii}^{aa}, the energy for each multiplet is minimized. The one-electron orbitals are determined.

Next, the CI (configuration interaction) calculations are performed, by using the MCSCF one-electron orbitals ψ_i's determined above, as a basis set and the lowest energy of each multiplet is obtained. Since a main part of the electron-correlation effect has already been included in determining the MCSCF one-electron orbitals, a small number of the Slater determinants are necessary in the CI calculations. Thus one can get a clear-cut-view of the many-electron states by this MCSCF-CI method, even when the correlation effect is strong. Thus the MCSCF-CI method is the most suitable variational method for a strongly correlated cluster system [130].

In the MCSCF method all the orbitals consisting of the Cu $3d_{x^2-y^2}$, $3d_{z^2}$, $4s$ and O $2p$ orbitals are taken into account in the summation over i and a in (5.1) and (5.2). In the CI calculation, all the single-electron excitation configurations among these orbitals are taken into account.

5.2 Choice of Basis Sets in the MCSCF-CI Calculations

We express the one-electron orbitals by linear combinations of atomic orbitals, where Cu $1s$, $2s$, $3s$, $4s$, $2p$, $3p$, $3d$ and O $1s$, $2s$, $2p$ orbitals are taken into account as the atomic orbitals. Each atomic orbital is represented by a linear combination of several Gaussian functions. For Cu $3d$, $4s$ and O $2s$, $2p$ atomic orbitals, we prepare two basis functions called "double zeta" for each orbital. Those are (12s6p4d)/[5s2p2d] for Cu [133] and (10s5p)/[3s2p] for O [134].

As for the oxygen ions, the diffuse components are usually used by researchers in the quantum chemistry. The diffuse components, however, cause problems for the point charge approximation outside of the cluster when a cluster is embedded in a crystal, because the diffuse components reach the nearest neighbour sites with considerable amplitudes. Instead of using the diffuse components for O^{2-}, Eto and Kamimura [103, 104] used extended O $2p$ basis functions which were originally prepared for a neutral atom, by introducing a scaling factor of 0.93. Then they multiplied all the Gaussian exponents in the double zeta base for the oxygen $2p$ orbitals by the same scaling factor of 0.93. This value of the scaling factor was determined so that the energy of an isolated O^{2-} ion should coincide with that obtained by the Hartree-Fock calculation.

5.3 Calculated Results of Hole-Doped CuO$_6$ Octahedrons in LSCO

As a model for cluster calculation by the MCSCF-CI method, we first choose a CuO$_6$ cluster in the LSCO compound as an object of study. Since the behaviours of planar and apical oxygen in a CuO$_6$ octahedrons in cuprates are very different, we investigate them in detail by labeling the oxygen in a CuO$_2$ plane as O(1) and the apical oxygen as O(2). As regards the lattice constants, we use them, as reported in [124], for LSCO.

Now we discuss the calculated results of a CuO$_6$ octahedron by the MCSCF-CI method. In the MCSCF-CI calculation Kamimura and Eto considered both the $^1A_{1g}$ and $^3B_{1g}$ multiplets independently. Then they compared the respective energies of both states to determine which is the ground state, varying the Cu–O(2) distance reflecting the anti-Jahn–Teller effect. In the following subsections we present the calculated results of $^1A_{1g}$ and $^3B_{1g}$ multiplets separately.

5.3.1 The $^1A_{1g}$ Multiplet (the Zhang–Rice Singlet)

The many-electron wavefunctions of the $^1A_{1g}$ multiplet are listed in Fig. 5.1, as a function of Cu-O(2) distance c. The sketch of one-electron orbitals which are obtained by the MCSCF method are shown in Fig. 5.2. As seen in Fig. 5.1, the many-electron wavefunction mainly consists of three electron configurations. In the first electron configuration at the left column in the figure, which has the largest coefficient, the Cu $d_{x^2-y^2}$-O(1) p_σ antibonding b_{1g} orbital, ψ_5, is unoccupied. In the second electron configuration at the center in the figure, the bonding orbital, ψ_4, is unoccupied while the antibonding

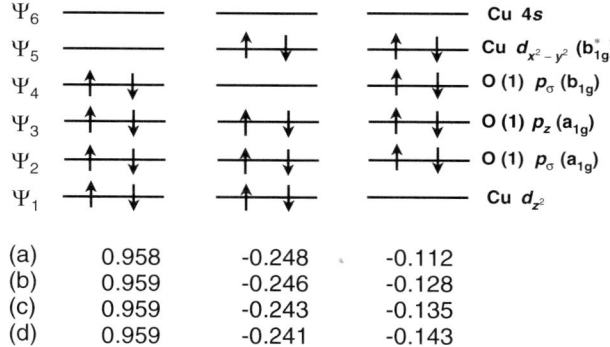

Fig. 5.1. The many-electron wavefunctions of $^1A_{1g}$ state in the hole-doped CuO$_6$ cluster. The Cu–O(2) distance, c, is (**a**) 2.41 Å, (**b**) 2.35 Å, (**c**) 2.30 Å and (**d**) 2.24 Å, respectively. The atomic orbital with the largest component in each MCSCF one-electron orbital is attached in the *right* side

40   5 MCSCF-CI Method

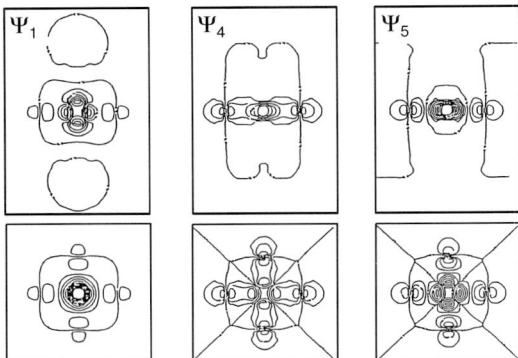

Fig. 5.2. The MCSCF one-electron orbitals optimized for $^1A_{1g}$ state in the hole-doped CuO_6 cluster ($c = 2.41$ Å). The *upper row* shows the wavefunctions perpendicular to the CuO_2 plane, while the *lower row* shows the wavefunctions in the CuO_2 plane. The contour lines are drawn every 0.05 a.u.

orbital, ψ_5, is doubly occupied. Thus the mixing between the first and the second configurations indicates that the holes occupy both of the Cu $d_{x^2-y^2}$ and the O(1) p_σ orbitals of b_{1g} symmetry and that a dopant hole forms a spin-singlet pair with the localized hole which occupies an antibonding b_{1g} orbital, b_{1g}^*. This situation corresponds to the Zhang–Rice singlet multiplet [96].

In the third electron configuration at the right column in Fig. 5.1, the a_{1g} orbital, ψ_1, is unoccupied while the b_{1g} orbitals, ψ_4 and ψ_5, are doubly occupied. The ψ_1, shown in Fig. 5.2, is almost localized at Cu d_{z^2}. This configuration appears for the following reason. When two holes are at a Cu site, the on-site Coulomb repulsion, the so-called Hubbard U, raises the energy. The Coulomb repulsion is smaller when the holes occupy both the d_{z^2} and the $d_{x^2-y^2}$ orbitals than when they remain only in the $d_{x^2-y^2}$ orbital. Thus the mixing of the $(d_{z^2})^2$ and the $(d_{x^2-y^2})^2$ electron configurations reduces the Hubbard U at the Cu site effectively, compared with the single configuration $(d_{x^2-y^2})^2$. This effect becomes larger as the Cu–O(2) distance decreases, as shown in Fig. 5.1.

5.3.2 The $^3B_{1g}$ Multiplet (the Hund's Coupling Triplet)

The $^3B_{1g}$ many-electron wavefunction is shown in Fig. 5.3. The a_{1g}^* orbital, ψ_4, and the b_{1g}^* orbital, ψ_5, are singly occupied and the two electrons couple to form a spin triplet by Hund's coupling. In ψ_5, $d_{x^2-y^2}$ is mixed with O(1) p_σ while ψ_4 consists almost entirely of d_{z^2}, as shown in Fig. 5.4. The strength of the on-site exchange energy, Hund's coupling, can be estimated from the energy difference between the $^3B_{1g}$ state and the excited $^1B_{1g}$ state. The estimated value is about 2.0 eV.

5.4 Z–R Singlet ($^1A_{1g}$) and Hund's Coupling Triplet ($^3B_{1g}$) Multiplets   41

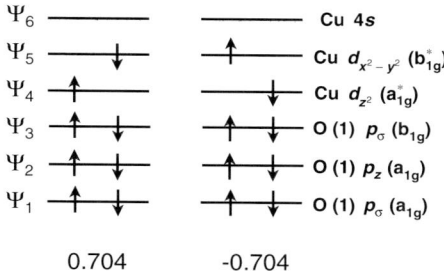

Fig. 5.3. The many-electron wavefunctions of $^3B_{1g}$ state in the hole-doped CuO_6 cluster ($c = 2.41$ Å)

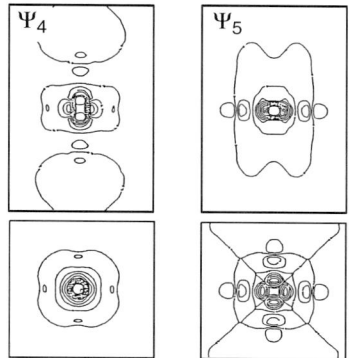

Fig. 5.4. The MCSCF one-electron orbitals optimized for $^3B_{1g}$ state in the hole-doped CuO_6 cluster ($c = 2.41$ Å). The *upper row* shows the wavefunctions perpendicular to the CuO_2 plane, while the *lower row* shows the wavefunctions in the CuO_2 plane. The contour lines are drawn every 0.05 a.u.

As the Cu–O(2) distance decreases and hence the CuO_6 cluster approaches a regular octahedron by doping, the $^1B_{1g}$ state becomes more stable. This is because the energy difference between the b_{1g}^* orbital and the a_{1g}^* orbital becomes smaller, so that the Hund's coupling becomes more effective.

5.4 Energy Difference between Zhang–Rice Singlet ($^1A_{1g}$) and Hund's Coupling Triplet ($^3B_{1g}$) Multiplets

The calculated energy difference between the $^1A_{1g}$ and the $^3B_{1g}$ states is shown in Fig. 5.5, as a function of the Cu–O(2) distance. The figure indicates that the ground state of the CuO_6 cluster changes from the $^1A_{1g}$ state to the $^3B_{1g}$ state when the Cu–O(2) distance decreases. The distance at which the transition occurs corresponds to the doping concentration $x \sim 0.1$ in the

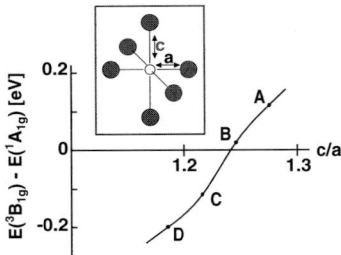

Fig. 5.5. The energy difference between the $^3B_{1g}$ and the $^1A_{1g}$ multiplets, as a function of the Cu–O(2) distance, c, in the hole-doped CuO_6 cluster. The Cu–O(1) distance, a, is fixed at 1.889 Å. The Cu–O(2) distance c is (A) 2.41 Å (undoped case), (B) 2.35 Å, (C) 2.30 Å ($La_{1.8}Sr_{0.2}CuO_4$) and (D) 2.24 Å, respectively

$La_{2-x}Sr_xCuO_4$ formula. Although the ground state of a CuO_6 octahedron embedded in LSCO changes from $^1A_{1g}$ to $^3B_{1g}$ multiplet by the anti-Jahn–Teller effect due to doping Sr^{2+} ions for La^{3+} ions, the energy difference between two multiplets in the underdoped hole-concentration regime is very small, i.e., at most 0.1 eV as seen in Fig. 5.5. Therefore, when the CuO_6 octahedrons form a CuO_2 network in LSCO, the transfer interaction between neighbouring octahedrons acts to mix two multiplets easily. This will lead to the Kamimura–Suwa model, as will be seen in Chap. 8.

6 Calculated Results of a Hole-Doped CuO$_5$ Pyramid in YBa$_2$Cu$_3$O$_{7-\delta}$

6.1 Introduction

In this chapter we discuss the results calculated with the MCSCF-CI method for a hole-doped CuO$_5$ pyramid in superconducting YBa$_2$Cu$_3$O$_7$ (abbreviated as YBCO$_7$) with $T_c = 90$ K and insulating YBa$_2$Cu$_3$O$_6$ (abbreviated as YBCO$_6$) performed by Kamimura and Sano [131]. The crystal structure of YBCO$_7$ is shown in Fig. 6.1(a). For comparison that of the insulating YBCO$_6$ is also shown in Fig. 6.1(b). A remarkable difference in the crystal structures

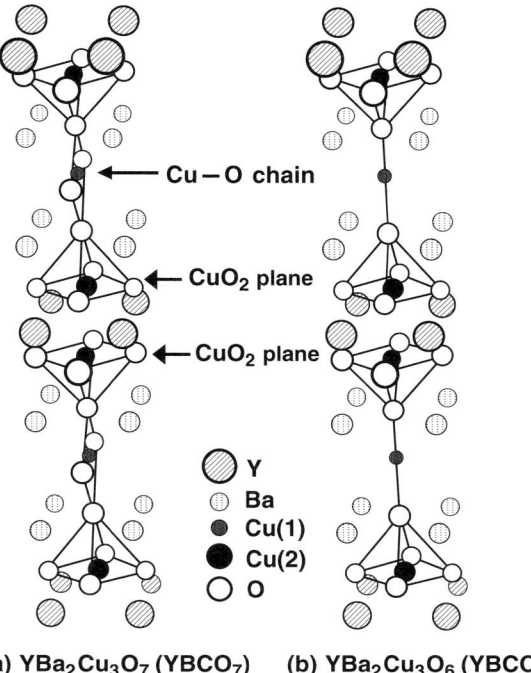

Fig. 6.1. The crystal structures of YBCO$_{7-\delta}$. (a) The orthorhombic structure of superconducting YBCO$_7$. (b) The tetragonal structure of insulating YBCO$_6$

of YBCO$_7$ and YBCO$_6$ is that there exists a Cu–O chain in YBCO$_7$. Like LSCO, there are two orbitals, an antibonding a_1 orbital, a_1^*, and a bonding b_1 orbital, b_1, as possible orbital states to accommodate dopant holes, where a point group of CuO$_5$ pyramid is C_{4v}. The sketch on the spatial extension of a_1^* and b_1 orbitals are shown in Fig. 4.4(a) and 4.4(b) in Chap. 4. As a result one has to deal with the $^1A_1$ and $^3B_1$ multiplets independently following the MCSCF-CI method. In doing so, we take into account the effect of Madelung potential from exterior ions outside the cluster by placing the point charge, +2 at Cu(2) in CuO$_2$ plane, +2 at Ba, +3 at Y, and −2 at O. As for the charge of Cu in the Cu–O chain (Cu(1)), q, we have taken $q = +1$ for insulating YBCO$_6$ from experimental (NMR) result [135]. This value is consistent with a condition of charge neutrality. However, in superconducting YBCO$_7$, the value of q is not clear. Thus Kamimura and Sano [131] calculated the energy difference between the $^1A_1$ and $^3B_1$ multiplets in the case of YBCO$_7$ as a function of q and then investigated the effect of inhomogeneous hole distribution in the Cu–O chain on the electronic state.

6.2 Energy Difference between $^1A_1$ and $^1B_1$ Multiplets

The calculated energy difference between the $^1A_1$ and the $^3B_1$ multiplets by Kamimura and Sano is shown in Fig. 6.2 as a function of the charge of Cu(1), q. The value of q and the existence of O^{2-} ions in a Cu–O chain play a crucial role in determining the Madelung energy at apical O site. There is an energy difference of 1.3 eV between $^3B_1$ and $^1A_1$ multiplets in insulating YBCO$_6$, as seen in Fig. 6.2 (closed circle), where the distance between Cu(2) and apical O, c, is taken as 2.47 Å. Cava et al. [107] observed the change of the apical O–Cu distance in YBCO as shown in Fig. 6.3 as a function of hole-concentration, where the apical O–Cu distance is denoted as Cu2-O1. One can see from this figure that the apical O–Cu distance in a CuO$_5$ pyramid in YBa$_2$CuO$_x$ decreases sharply from 2.44 Å to 2.29 Å as the oxygen content x changes from 6.4 to 7.0, where YBCO$_x$ shows the highest T_c of 90 K with $x = 7.0$.

In the case of insulating YBCO$_6$, the energy difference between $^3B_1$ and $^1A_1$ multiplets is 1.3 eV, as shown by the closed circles in Fig. 6.2, where the distance between Cu(2) and apical O ions is fixed at 2.47 Å. In Fig. 6.2 the open circles show the energy difference for superconducting YBCO$_7$ as a function of q, where c is fixed at 2.29 Å and oxygen atoms are introduced into a Cu–O chain. It is clear from this figure that, when the value of q decreases, the ground state of the CuO$_5$ pyramid in YBCO$_7$ changes from the $^1A_1$ to $^3B_1$ around the $q \approx 1.45$. This is because, as the value of q decreases and thus the Maderung potential at the apical oxygen site decreases, the energy difference between the a_1^* orbital which contains the p_z orbital at apical oxygen site and the b_1 orbital becomes smaller, so that the role of Hund's coupling becomes more effective.

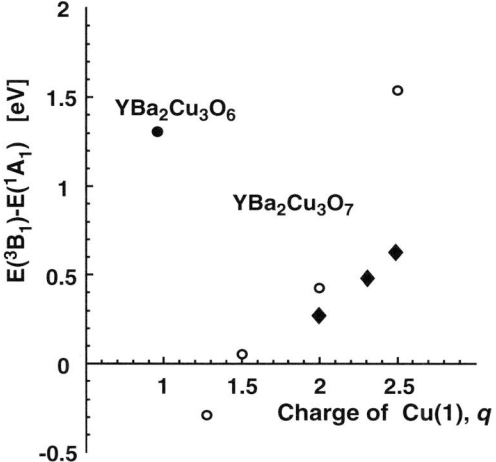

Fig. 6.2. The energy difference between the $^3B_1$ and the $^1A_1$ multiplets, as a function of the charge of a Cu(1) ion in the Cu–O chain, q, in the hole-doped CuO$_5$ cluster embedded in YBCO$_6$ and YBCO$_7$. The *closed circle* represents the energy difference between the $^3B_1$ and the $^1A_1$ multiplets in insulating YBCO$_6$ [136, 137]. The *open circles* represent the energy difference between the $^3B_1$ and the $^1A_1$ multiplets in superconducting YBCO$_7$ as a function of constant q for all Cu(1) ions, where c is fixed at 2.29 Å. Further the *solid diamonds* represent the calculated results in the case of CDW in a Cu–O chain

6.3 Effect of Change Density Wave (CDW) in a Cu–O Chain

In a previous section we saw that in superconducting YBCO$_7$ the calculated lowest state energy is very sensitive to the charge of Cu(1) in the Cu–O chain, q. In this section we discuss how the multiplets of a CuO$_5$ pyramid are affected by the inhomogeneous hole distribution in a Cu–O chain, that is the charge density wave (CDW), based on the calculated results by Kamimura and Sano [131]. The existence of such CDW in a Cu–O chain in YBCO$_7$ was reported by various experimental groups. For example, a scanning tunneling microscopy (STM) experiment [138], neutron inelastic scattering experiments [139] and diffuse X-ray scattering [140] have reported on the existence of CDW in a Cu–O chain in YBCO$_7$. In this context Kamimura and Sano [131] tried to clarify theoretically how the CDW in the Cu–O chains affect the electronic structure of a hole-doped CuO$_5$ pyramid. This was the first theoretical study on the CDW effect in a Cu–O chain on the electronic structure of a CuO$_5$ pyramid in YBCO$_7$. Following Kamimura and Sano [131], let us explain how the CDW in a Cu–O chain influences the electronic structure of a CuO$_5$ pyramid.

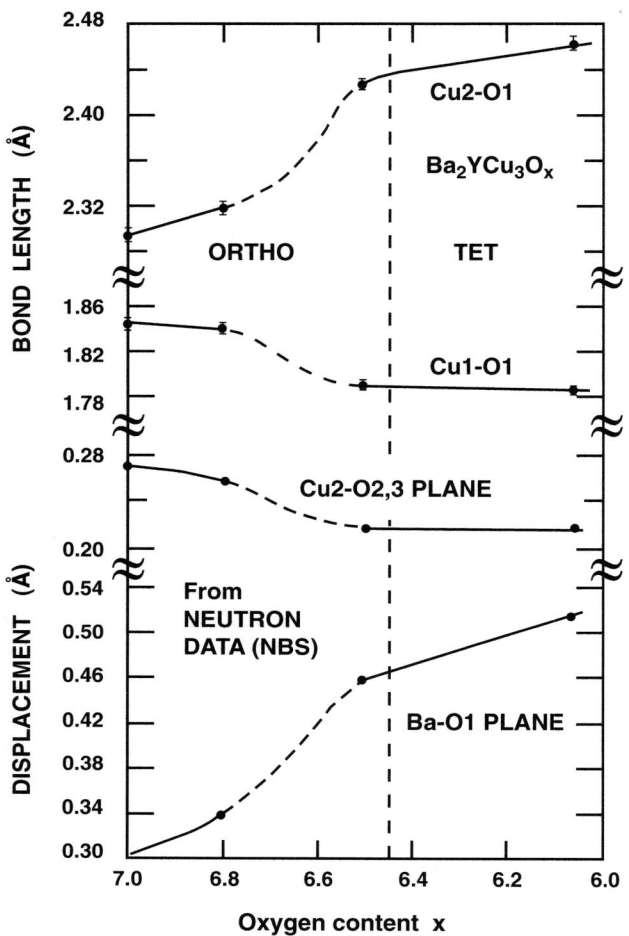

Fig. 6.3. The variation of the lattice parameters with oxygen content x in the YBa$_2$Cu$_3$O$_x$ (after [107])

Suppose that the charge of a Cu(1), q, is +2.5 and that the states of holes in a Cu–O chain are expressed by a one-dimensional energy band. This means that there are 1.5 holes in a Cu–O chain and that three quarters of the energy band for a Cu–O chain are filled by holes. In this case the Fermi wavenumber k_F is given by $\pi/4a$ approximately, where a is a Cu(1)–Cu(1) distance along the chain and it is 3.8 Å for YBCO$_7$. Thus the CDW modulation-wavelength becomes 15.2 Å, because the modulation wavelength λ_{CDW} is given by $\lambda_{CDW} = 2\pi/2k_F$ and it is nearly equal to $4a$. This value is consistent with the experimental results [138, 139], since the observed modulation wavelength of CDW in a Cu–O chain takes a value between 13 ∼ 16 Å.

6.3 Effect of Change Density Wave (CDW) in a Cu–O Chain

In superconducting YBCO$_7$ an oxygen introduced in a Cu–O chain produces two holes in a unit cell consisting of a Cu–O chain and two CuO$_2$ planes. Considering the charge of +3e for Y, +2e for Ba, +2e for Cu(2) and −2e for O in CuO$_2$ plane, +1e for Cu(1) and −2e for O in a Cu–O chain, and further distributing the charge of dopant holes over both a Cu–O chain and two CuO$_2$ planes in a unit cell, the following equation holds for a relation between the number of holes in Cu–O chain, η, and that of a CuO$_2$ plane, ζ, in the unit cell from the condition of the charge neutrality;

$$\eta + 2\zeta = 2 . \tag{6.1}$$

Then the charge of Cu in a Cu–O chain, q, is related to η by the relation $q = 1 + \eta$. Since the values of η and ζ have not been determined experimentally so far, Kamimura and Sano calculated the lowest energies of the $^1A_1$ and $^3B_1$ multiplets by varying a value of η. In the case of the uniform charge distribution for the charge of Cu(1) in a Cu–O chain, for example, for $q = +2.5$, η becomes 1.5 and thus ζ is 0.25 from (6.1). This means that one hole exists per four CuO$_5$ pyramids. Since a CuO$_5$ pyramid is embedded in YBCO$_7$, one must take into account the effect of Madelung potential from exterior ions outside the pyramid by putting the point charge +2e at Cu(2) in CuO$_2$ plane, +2e at Ba, +3e at Y, −2e at O. As to the charge of Cu(1) in a Cu–O chain, one may place the point charges according to the CDW modulation-wavelength, as shown on line A in Fig. 6.4. For example, Cu(1)$^{+1.75}$ ions are placed at the interval of every four Cu(1) sites along the line of the Cu–O chain. This corresponds to the case that the Cu(1) right above the CuO$_5$ pyramid under consideration has the charge of +1.75e, while the charge of +2.75e is placed at remaining Cu(1) sites on the line A. Thus the averaged charge of Cu(1) atoms on the line A is +2.5e. In the same way one may put the charge of +3e at Cu(2) sites at the interval of four sites with the same modulation as that of the Cu–O chain and put the charge of +2e at the remaining Cu(2) sites on the line B as seen in Fig. 6.4. The line B includes the CuO$_5$ pyramid under consideration. Thus the averaged charge of the Cu(2) ions and the averaged hole concentration in a CuO$_2$ plane on the line B becomes 2.25e and 0.25, respectively. As to the charges of all the Cu(2) ions except those on the line B, one can take +2.25e as an averaged charge, while as regards the charges of all the Cu(1) ions except the Cu(1) ions on the line A, one can take +2.5e as an averaged charge, as shown in Fig. 6.4. In this way the CDW-like hole distribution is formed under the condition in which the charge neutrality is kept.

On the basis of the charge distribution shown in Fig. 6.4, Kamimura and Sano have calculated the lowest energies of the $^1A_1$ and $^3B_1$ multiplets by the MCSCF-CI method. The calculated results are shown by solid diamonds in Fig. 6.2, where the energy difference between the $^1A_1$ and $^3B_1$ multiplets is shown on the vertical axis and $\bar{q}$ on the horizontal line represents the averaged charge of Cu(1) ions in a Cu–O chain. For comparison, we also show by open

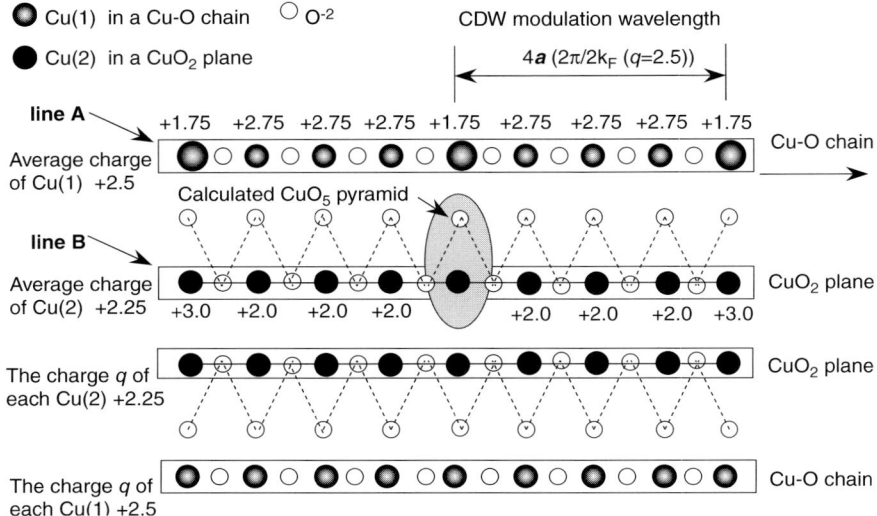

Fig. 6.4. The charge distribution of Cu(1) ions in Cu–O chains and of Cu(2) ions in CuO$_2$ planes for the case in which the charge of Cu(1) is modulated by the CDW modulation wavelength and the average value of Cu(1)'s charge, $\bar{q}$, is equal to 2.5. The line A represents the Cu–O chain which includes the Cu(1) ion right above the hatched CuO$_5$ pyramid under consideration. The electronic structure of a CuO$_5$ pyramid marked by hatch on line B is calculated by the MCSCF-CI method in the presence of the CDW charge distribution shown in this figure, whose effect is considered as the Madelung potential in the MCSCF-CI calculations

circles the energy difference between the $^1A_1$ and $^3B_1$ multiplets calculated for the case of the constant charge distribution in a Cu–O chain as a function of q [136, 137].

As shown in Fig. 6.2, in the case of the constant charge distribution of the Cu(1) ions in Cu–O chain, the energy difference between $^1A_1$ and $^3B_1$ multiplets is larger than that in the case of CDW. For example, the former is 1.55 eV for $q = +2.5$. In the CDW case shown by solid diamonds in Fig. 6.2, the calculated energy difference between the $^1A_1$ and the $^3B_1$ multiplets is significantly reduced. For example, in the case of $\bar{q} = +2.5$ it becomes 0.65 eV. Thus the electronic structure is strongly affected by the charge distribution in a Cu–O chain caused by CDW. The decrease of the energy difference between these two multiplets is reasonable because in this case the Madelung potential at the apical O in a CuO$_5$ pyramid becomes lower for hole carriers. However, since the holes in Cu–O chains occupy both Cu(1) and O sites, the charge of Cu(1) in a Cu–O chain becomes lower than +2.5e. This favors the $^3B_1$ multiplet energetically, because the Madelung potential at the apical O site becomes further lower for a hole carrier so that a probability of occupying the apical O site increases for the hole carrier. In the case of YBCO, the $^1A_1$

6.3 Effect of Change Density Wave (CDW) in a Cu–O Chain

multiplet is always lower in its ground state energy than the $^3B_1$ multiplet. Thus we conclude that, when the averaged charge of Cu(1) ions takes a value between 2.0 and 2.3, the energy difference between the $^1A_1$ and $^3B_1$ multiplets becomes of the same order of magnitudes as transfer interaction between $^3B_1$ and $^1A_1$ multiplets at neighbouring CuO_5 pyramids, 0.4 eV, by the existence of CDW in a Cu–O chain. In this context a hole carrier can hop between $^1A_1$ and $^3B_1$ multiplets on the neighbouring CuO_5 pyramids, when the CDW exists on the Cu–O chains. This makes the existence of the Kamimura–Suwa model possible.

7 Electronic Structure of a CuO$_5$ Pyramid in Bi$_2$Sr$_2$CaCu$_2$O$_{8+\delta}$

7.1 Introduction

In this chapter we discuss the calculated results for the electronic structures of a CuO$_5$ pyramid embedded in Bi$_2$Sr$_2$CaCu$_2$O$_{8+\delta}$ with use of the MCSCF-CI method by Tobita and Kamimura [132]. The high T_c superconductors of the Bi–Sr–Ca–Cu–O materials system were discovered by Maeda et al. in 1988 [5]. The composition of these materials is determined as Bi$_2$Sr$_2$Ca$_{n-1}$Cu$_n$O$_{4+2n+\delta}$ with n being 1, 2, and 3. To distinguish the values of different n, these compounds are distinguished as Bi2201 ($n = 1$), Bi2212 ($n = 2$) and Bi2223 ($n = 3$), where T_c of Bi2201, Bi2212 and Bi2223 are 20 and 80, 110 K, respectively. The number of the CuO$_2$ planes increases with increasing n. These compounds have the Bi$_2$O$_2$ blocking layers. In the chemical formula of "Bi$_2$Sr$_2$CaCu$_2$O$_{8+\delta}$", δ represents the excess of oxygen. When excess oxygen does not exist, i.e. $\delta = 0$, this material is an insulator. When excess oxygen is introduced, hole carriers are supplied into the CuO$_2$ planes, and this material shows superconductivity. When increasing the value of δ, T_c of Bi2212 rises. Thus many researchers regard the excess of oxygen as an origin of carriers which are responsible for superconductivity.

7.2 Models for Calculations

Figures 7.1(a) and (b) show the crystal structures of Bi2212 for $\delta = 0$ and $\delta = 0.25$, respectively. In these structures, the distance between Cu and apical O in the CuO$_5$ pyramid cluster is 2.15 Å for $\delta = 0.25$ [127]. It is very short compared with the distance in insulating Bi2212 with $\delta = 0$ which is 2.47 Å. Thus the local distortion of CuO$_5$ pyramids is expected to play an important role in determining their electronic structure.

We consider that the case of $\delta = 0.25$ corresponds to the optimum doping in Bi2212. In this section, we pay attention to the electronic structures of a CuO$_5$ pyramid in the cases of $\delta = 0$ and $\delta = 0.25$. According to the observation by transmission electron microscope (TEM) [141], the Bi$_2$O$_2$ blocking layers are slightly distorted from the crystal structures shown in Figs. 7.1(a) and (b), and undulation appears along the b axis. Thus, a real crystal structure of Bi2212 is more complex than the structures shown in Figs. 7.1(a) and

52   7 Electronic Structure of a CuO$_5$ Pyramid in Bi$_2$Sr$_2$CaCu$_2$O$_{8+\delta}$

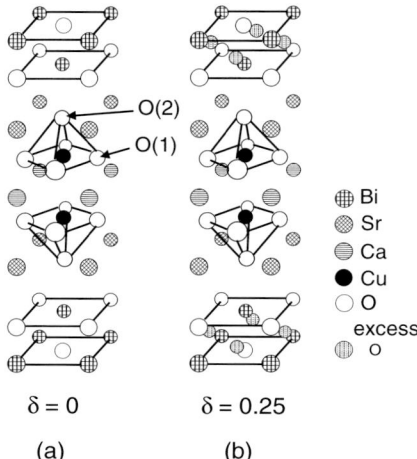

Fig. 7.1. The crystal structures of Bi$_2$Sr$_2$CaCu$_2$O$_{8+\delta}$. Here, (a) and (b) represent the structures for $\delta = 0$ [5] and 0.25, respectively

(b). The origin of this distorted structure may be considered for the following reasons: The excess oxygen enters into the middle of the Bi$_2$O$_2$ blocking layers. Depending on whether the Bi$_2$O$_2$ blocking layers include the excess oxygen or not, the Bi$_2$O$_2$ blocking layers show a slightly irregular structure. However, the hole carriers cannot recognize such a slight change of the structure, because its mean free path is much longer than Cu–O–Cu distance. In this context Tobita and Kamimura [132] used the average structures shown in Figs. 7.1(a) and (b) for the calculation of electronic structures. Further, in the case of $\delta = 0.25$, the excess oxygen of charge $-0.5e$ are placed at four sites in every middle region between the Bi$_2$O$_2$ blocking layers, because the hole carriers are subject to the average Madelung potential from the excess oxygen of charge $-2e$, which are distributed randomly between the Bi$_2$O$_2$ blocking layers. We call the crystal structure shown in Fig. 7.1(b) a "virtual crystal structure" in this respect.

In calculations by Tobita and Kamimura, 742 and 846 ions outside the CuO$_5$ pyramid under consideration are treated as point charges to consider the effect of Madelung potential for the case of $\delta = 0$ and $\delta = 0.25$, respectively. Then they calculated the electric structures of a single CuO$_5$ pyramid using the crystal structures shown in Figs. 7.1(a) and (b) for the case of $\delta = 0$ and $\delta = 0.25$, respectively.

7.3 Calculated Results

The calculated energy difference between the $^1A_1$ and $^3B_1$ states is about 2.15 eV for the case of $\delta = 0$. The energy of the $^1A_1$ state is lower than that of the $^3B_1$ state. Since the transfer interaction between neighbouring pyramids is about 0.4 eV, a dopant hole is localized around a particular CuO$_5$ pyramid in the case of $\delta = 0$. As a result Bi2212 with $\delta = 0$ is an insulator, consistent with experimental results [5, 127].

For the case of $\delta = 0.25$, on the other hand, the energy difference between the $^1A_1$ and $^3B_1$ states is about 0.034 eV. The energy of the $^1A_1$ multiplet is still lower than that of the $^3B_1$ multiplet. Since this energy difference is very small compared with the transfer interaction between the b_1 and a_1^* orbitals in the neighbouring CuO$_5$ pyramids, which is about 0.4 eV, two states are mixed by the transfer interaction between the neighbouring CuO$_5$ pyramids, and a coherent state is expected to be composed in a superconducting Bi2212 material, when the localized spin forms an antiferromagnetic order in a spin-correlated region, as will be described in the following chapter.

A reason why the difference between the $^1A_1$ and $^3B_1$ states decreases is the following: As the value of δ increases, the Madelung potential at an apical oxygen site decreases. As a result the energy difference between the energy of the a_1^* orbital, which contains the p_z orbital at the apical oxygen site, and that of the b_1 orbital becomes smaller, so that the Hund's coupling becomes more effective. Thus, the energy difference between the $^1A_1$ and $^3B_1$ states becomes smaller.

7.4 Remarks on Cuprates in which the Cu–Apical O Distance is Large

As we have described in this chapter, in the case of Bi–Sr–Ca–Cu–O materials system, the dopant holes are provided from the excess oxygen in the Bi$_2$O$_2$ blocking layers. A similar situation appears in other cuprates such as TlBa$_2$CaCu$_2$O$_7$ (Tl1212) with $T_c = 103$ K [6] and HgBa$_2$Ca$_2$Cu$_3$O$_{8+\delta}$ (Hg1223) $T_c = 135$ K [8]. In these cases the blocking layers are negatively charged after dopant holes are provided into the CuO$_2$ layers. Thus an existence probability of hole-carriers at apical oxygen sites is not low, even when the apical O–Cu distance is very large such as 2.76 Å and 2.74 Å in Tl1212 and Hg1223, respectively, because the attractive electrostatic interaction favors the hopping of hole-carriers to the apical oxygen sites, which are close to the blocking layers. Thus the Kamimura–Suwa model holds even when the apical O–Cu distance is large, although the mixing ratio of the Hund's coupling triplet into the Zhang–Rice singlet is small, compared with the cases of LSCO, YBCO$_7$, Bi2212 materials, in which the apical O–Cu distance is shorter than 2.41 Å, the length between apical O and Cu ions in the Jahn–Teller elongated octahedron or pyramid.

Further, in Tl1212 and Hg1223, the length of the c-axis is much longer than those of LSCO, YBCO$_7$, Bi2212. This indicates that the two-dimensional nature is stronger in Tl1212 and Hg1223, favoring the occurrence of higher values of T_c in superconductivity.

8 The Kamimura–Suwa (K–S) Model: Electronic Structure of Underdoped Cuprates

8.1 Description of the Model

Now we construct the many-electron electronic structure of underdoped cuprates, based on the calculated results of a CuO_6 octahedron embedded in LSCO and of a CuO_5 pyramid in $YBCO_7$ and Bi2212 described in Chaps. 4 to 7. Before presenting results, we briefly describe the theoretical treatment made by Kamiumra and Suwa [15], which is now called the Kamimura–Suwa (K–S) model. As an example, we choose LSCO here. According to the Kamimura–Suwa model, there exist areas in each CuO_2 layer in which the localized spins form the antiferromagnetic (AF) order. Here we call these areas "spin-correlated regions". The size of each spin-correlated region is characterized by the spin-correlation length. Then, following the results of Kamimura and Eto [104], a dopant hole with up spin in a spin-correlated region occupies an a_{1g}^* orbital, $\phi_{a_{1g}^*}$, at CuO_6's with localized up-spins, because of an energy gain of about 2 eV due to the intra-atomic exchange interaction between the spins of an a_{1g}^* hole and of a localized hole in an antibonding b_{1g} orbital (b_{1g}^*) (Hund's coupling) within the same CuO_6 octahedron, as shown in Fig. 8.1(c). As a result the spin-triplet $^3B_{1g}$ multiplet is created. Since Hund's coupling prevents a hole with up spin from occupying an a_{1g}^* orbital in a CuO_6 octahedron with a localized down-spin, a hole with up-spin in a CuO_6 octahedron with a localized up-spin can not hop into neighbouring a_{1g}^* orbital. Instead, it can enter into a bonding b_{1g} orbital, $\phi_{b_{1g}}$, in a neighbouring CuO_6 octahedron with a localized down-spin without destroying the antiferromagnetic order. In this case there is the energy gain of about 4.0 eV due to the antiferromagnetic exchange interaction between holes in bonding and antibonding b_{1g} orbitals, as shown in Fig. 8.1(c). This results in the Zhang–Rice singlet state $^1A_{1g}$.

By taking account of the energy difference between the a_{1g}^* and b_{1g} orbitals, the energy difference between the highest occupied states in the $^3B_{1g}$ and $^1A_{1g}$ multiplets becomes of the same order of magnitudes as a transfer interaction between a_{1g}^* and b_{1g} orbitals in the neighbouring CuO_6 octahedrons. In this way the dopant holes can move resonantly from a CuO_6 to a neighbouring CuO_6 in a CuO_2 layer by the transfer interaction of about 0.3 eV without destroying the local antiferromagnetic (AF) order, as shown in Figs. 8.1(a) and (b). Such coherent motion of the dopant holes is possible

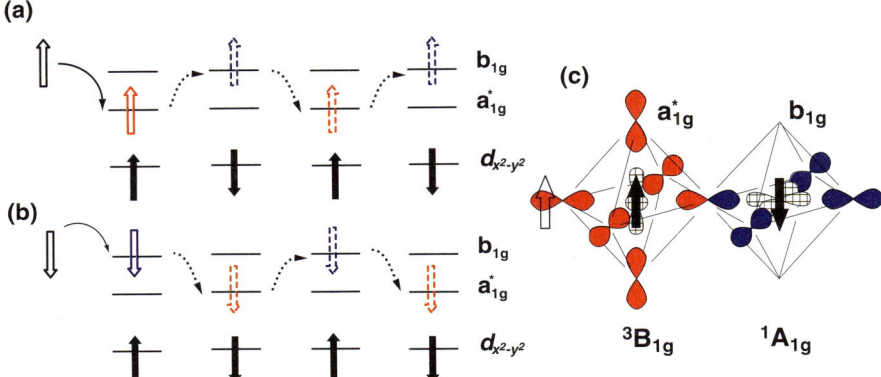

Fig. 8.1. Schematic view of the K–S model in LSCO, representing the coherent motion of a dopant hole from $^3B_{1g}$ multiplet (a^*_{1g} orbital state) to $^1A_{1g}$ multiplet (b_{1g} orbital state) in the presence of the AF ordering of the localized spin system. Here (**a**) and (**b**) correspond to a coherent motion of up-spin and down-spin states of dopant holes, respectively, where a dopant hole of a *solid white arrow* at the *left-hand side* in (**a**) and (**b**) moves to positions of *dotted white arrows* in turn by the transfer interactions. Figure (**c**) represents a coherent motion of an up-spin carrier from $^3B_{1g}$ to $^1A_{1g}$ multiplet. It should be noticed that the relative position of a^*_{1g} and b_{1g} levels changes according to the doping concentration. The energy levels in this figure are obtained from the results of Kamimura and Eto ([104])

when the spin-correlation length is much larger than the distance between neighbouring copper sites and the magnitudes of transfer interactions between neighbouring CuO_6 octahedrons are comparable to the energy difference between the highest occupied orbital states in the Zhang–Rice singlet and the Hund's coupling triplet. As a result a metallic state is created due to the delocalization effect of the dopant hole, and it simultaneously causes d-wave superconductivity when the temperature is below T_c, as was shown by Kamimura et al. [30]. We will describe d-wave superconductivity in Chap. 14.

Kamimura and Suwa [15] expressed the above coherent motion of dopant holes with up and down spins in a metallic state with the following forms of Bloch-type wave functions:

$$\Psi_{k\alpha}(r)\chi = \sum_{R} \exp(ik \cdot R)\left[A_k \phi_{a^*_{1g}}(r - R) + B_k \phi_{b_{1g}}(r - R - a)\right]\alpha\chi \tag{8.1}$$

and

$$\Psi_{k\beta}(r)\chi = \sum_{R} \exp(ik \cdot R)\left[A_k \phi_{a^*_{1g}}(r - R - a) + B_k \phi_{b_{1g}}(r - R)\right]\beta\chi \tag{8.2}$$

where α and β represent the up- and down-spin states of a dopant hole, respectively. Further, the spin function χ represents the antiferromagnetic ordering state of the Cu localized spins in a CuO$_2$ layer, where the up and down localized spins are assigned at $\boldsymbol{R}$ and $\boldsymbol{R}+\boldsymbol{a}$ Cu sites, respectively. Furthermore, $\boldsymbol{a}$ is a vector representing the distance between Cu sites with localized up and down spins in an antiferromagnetic unit cell. The summation over $\boldsymbol{R}$ is taken for the antiferromagnetic unit cells. In both (8.1) and (8.2), the first and the second terms in the square brackets represent the Hund's coupling and Zhang–Rice multiplets, respectively. In the case of YBCO$_7$, the coherent motion of a dopant hole due to the alternate appearance of the $^1A_1$ and $^3B_1$ multiplets is also possible when the CDW exists in a Cu–O chain as described in Chap. 6, and in Bi2212 the coherent motion always occurs for $\delta = 0.25$, as shown in Chap. 7.

A schematic picture of the K–S model in YBCO$_7$ and Bi2212 representing the coherent motion of a dopant hole with up and down spins in the presence of local AF order is shown in Fig. 8.2. As mentioned in Sect. 4.4, the distance between apical oxygen and Cu in CuO$_6$ octahedrons in LSCO [108, 123] or CuO$_5$ pyramids in YBCO [107, 109] and Bi2212 [5, 127] becomes shorter due to the anti-Jahn–Teller effect when the hole-concentration changes from an insulating phase to a superconducting phase. In order for the K–S model to hold, the effects of the local lattice distortions due to the anti-Jahn–Teller effect are essentially important.

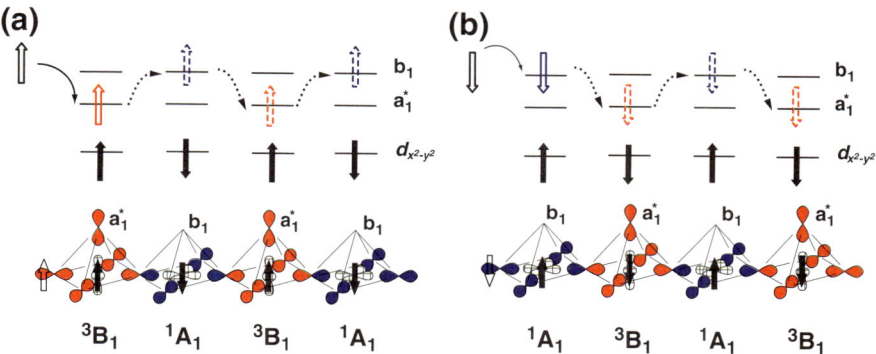

Fig. 8.2. Schematic view of the K–S model in YBCO$_7$ and Bi2212, representing the coherent motion of a dopant hole from $^3B_1$ multiplet (a$^*_{1g}$ orbital state) to $^1A_1$ multiplet (b$_{1g}$ orbital state) in the presence of the AF ordering of the localized spin system. Here (**a**) and (**b**) correspond to up-spin and down-spin states of dopant holes, respectively, where a dopant hole of a *solid white arrow* at the *left-hand side* in (**a**) and (**b**) moves to positions of *dotted white arrows* in turn by the transfer interactions

8.2 Experimental Evidence in Support of the K–S Model

8.2.1 Existence of the Antiferromagnetic Spin Correlation in the Underdoped Regime

In the Kamimura–Suwa model the spin-correlation length must increase in the underdoped region when the hole concentration increases, in order for every hole-carrier to move over a considerable distance without interacting with each other. As the result of the delocalization effect of hole carriers, a metallic state is created. As to the hole-concentration dependence of the spin correlation length, Mason et al. [142], Yamada et al. [143] and Lee et al. [53] reported that the spin-correlation length in the underdoped region of $La_{2-x}Sr_xCuO_4$ increases from $x = 0.05$, the onset of superconductivity, with increase of hole concentration x and reaches a value of more than 50 Å for the optimum doping ($x = 0.15$). These experimental results support the K–S model in which a metallic and superconducting state corresponds to a coherent state characterized by the coexistence of the local AF ordering and of the ordering with regard to the alternating appearance of the $^3B_{1g}$ and $^1A_{1g}$ multiplets in the carrier system.

8.2.2 Coexistence of the $^1A_{1g}$ and the $^3B_{1g}$ Multiplets

In order to investigate the coexistence of the $^3B_{1g}$ and $^1A_{1g}$ multiplets in the underdoped regime, Chen et al. [93] performed polarization-dependent X-ray absorption measurements for O K and Cu L edges in LSCO. For the Cu L edge, they observed a doping-induced satellite peak (L_3') for both polarizations of the electric vector of the X-rays $\boldsymbol{E}$, parallel and perpendicular to the c-axis, in a shoulder area of the doping-independent Cu L_3 line, with an intensity ratio of about 1 to 9, where a main L_3 line corresponds to transitions from a Cu 2p core level to the upper Hubbard Cu $d_{x^2-y^2}$ band, indicating the existence of the localized spins. Since the former ($\boldsymbol{E} \parallel c$) and the latter ($\boldsymbol{E} \perp c$) polarizations detect the Hund's coupling triplet ($^3B_{1g}$) and the Zhang–Rice singlet ($^1A_{1g}$), respectively, the appearance of the doping-induced satellite peak for both polarizations at the same energy suggests that the state of the dopant holes must be a single coherent state consisting of the Hund's coupling triplet multiplet and the Zhang–Rice singlet multiplet. For $Tl_2Ba_2CaCu_2O_8$ and $Tl_2Ba_2Ca_2Cu_3O_3$ as well as LSCO, Pellegrin et al. [94] also found the polarization dependence similar to that found by Chen et al. for LSCO [93].

In 1989 Bianconi et al. [144] also reported that the peak energy separation between transitions for polarizations parallel and perpendicular to the c-axis in LSCO decreases towards zero when the Sr concentration increases from a non-superconducting regime to a superconducting regime, consistent with the above experimental results. The existence of localized spins on Cu indicated by the observation of the Cu L_3 line is also supported by neutron scattering experiments. For example, Birgeneau et al. [145] showed the coexistence of the spin-correlation of localized spins in the AF order and superconductivity in LSCO; that is, the spins of Cu $d_{x^2-y^2}$ holes form a two-dimensional (2D) local antiferromagnetic (AF) order even in the superconducting state.

Recently the site-specific X-ray absorption spectroscopy of $YBa_2Cu_3O_{6.91}$ with $T_c = 92$ K by Merz et al. [111] determined the hole distribution in a CuO_2 plane, at an apical O site and in a Cu–O chain. According to this result the experimental values of hole distribution in a CuO_2 plane, at an apical O site and in a Cu-O chain are 0.40, 0.27 and 0.24, respectively. These results are consistent with the theoretical values calculated by Kamimura and Sano [131]. In particular, this experimental result clarified an important role of an apical oxygen site in a CuO_5 pyramid in the electronic structure of superconducting $YBCO_{6.91}$. This is an important experimental evidence for the K–S model.

8.3 Hamiltonian for the Kamimura–Suwa Model (The K–S Hamiltonian)

Kamimura and Suwa introduced the following effective Hamiltonian H_{KS} in order to describe the K–S model. As seen below, it consists of five terms: the effective one-electron Hamiltonian H_{eff} for $a_{1g}^*(a_1^*)$ and $b_{1g}(b_1)$ orbital states, the transfer interaction between neighbouring CuO_6 octahedrons (CuO_5 pyramids) H_{tr}, the superexchange interaction between the Cu $d_{x^2-y^2}$ localized spins H_{AF}, and the exchange interactions between the spins of dopant holes and $d_{x^2-y^2}$ localized holes within the same CuO_6 octahedron (CuO_5 pyramid) H_{ex}, and the repulsive interaction between dopant holes on $a_{1g}^*(a_1^*)$ or $b_{1g}(b_1)$ orbital within the same CuO_6 octahedron (CuO_5 pyramid) H_U. Thus we have

$$H_{KS} = H_{\text{eff}} + H_{\text{tr}} + H_{\text{AF}} + H_{\text{ex}} + H_U$$
$$= \sum_{i,m,\sigma} \varepsilon_m C_{im\sigma}^\dagger C_{im\sigma} + \sum_{\langle i,j\rangle,m,n,\sigma} t_{mn} \left(C_{im\sigma}^\dagger C_{jn\sigma} + \text{h.c.} \right)$$
$$+ J \sum_{\langle i,j\rangle} \boldsymbol{S}_i \cdot \boldsymbol{S}_j + \sum_{i,m} K_m \, \boldsymbol{s}_{i,m} \cdot \boldsymbol{S}_i + U \sum_{i,m} \hat{n}_{i,m,\uparrow} \hat{n}_{i,m,\downarrow}, \quad (8.3)$$

where ε_m ($m = a_{1g}^*(a_1^*)$ or $b_{1g}(b_1)$) represents the effective one-electron energy of the $a_{1g}^*(a_1^*)$ and $b_{1g}(b_1)$ orbitals, $C_{im\sigma}^\dagger$ and $C_{im\sigma}$ the creation and annihilation operators of a dopant hole in m-type orbital with spin σ in the ith CuO_6 octahedron or ith CuO_5 pyramid, respectively, t_{mn} the effective transfer integrals of a dopant hole between m-type and n-type orbitals of neighbouring CuO_6 octahedrons (CuO_5 pyramids), J the superexchange coupling between the spins $\boldsymbol{S}_i$ and $\boldsymbol{S}_j$ of $d_{x^2-y^2}$ localized holes in the b_{1g}^* (b_1^*) orbital

8 The Kamimura–Suwa (K–S) Model

at the nearest neighbour Cu sites i and j ($J > 0$ for AF interaction), K_m the exchange integral between the spin of a dopant hole $s_{i,m}$ and a $d_{x^2-y^2}$ localized spin S_i in the ith CuO$_6$ octahedron or ith CuO$_5$ pyramid ($K_{a_{1g}^*}(K_{a_1^*})$ < 0 for the Hund's coupling spin triplet multiplet with a$_{1g}^*$(a$_1^*$) orbital state and $K_{b_{1g}}(K_{b_1}) > 0$ for the Zhang–Rice spin singlet multiplet with b$_{1g}$(b$_1$) orbital state), and U the Hubbard U-like interaction with $\hat{n}_{i,m,\sigma} = C_{im\sigma}^\dagger C_{im\sigma}$, where $\hat{n}_{i,m,\sigma}$ is a number operator. Here $s_{i,m} = \sum_{\sigma\sigma'} C_{im\sigma}^\dagger \sigma_{\sigma\sigma'} C_{im\sigma'}$ with Pauli matrices $\sigma_{\sigma\sigma'}$. When the total number of dopant holes in a system consisting of L sites is denoted as N, the following relation holds;

$$\sum_{i,m,\sigma} \langle \hat{n}_{i,m,\sigma} \rangle = N \ . \tag{8.4}$$

Hereafter we call the effective Hamiltonian of the K–S model (8.3) the K–S Hamiltonian. The role of each term in the K–S Hamiltonian (8.3) is schematically shown in Fig. 8.3.

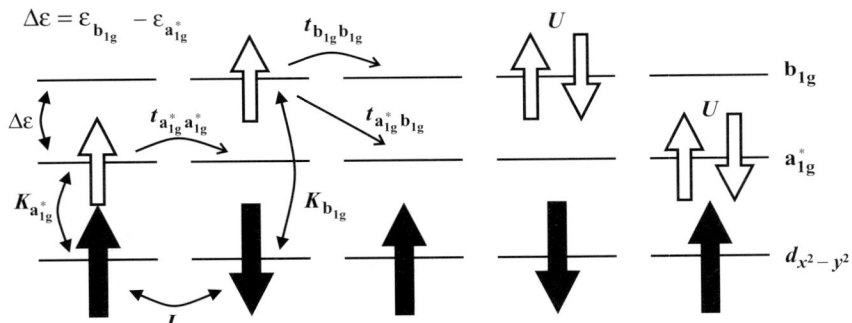

Fig. 8.3. Schematic explanation of the various interactions in the K–S Hamiltonian. *Black arrows* represent the localized spin at Cu site, while *white arrows* represent the spins of dopant holes

As regards the parameters in the K–S Hamiltonian (8.3), we adopt the following values; $J = 0.1$ from the experiment of the magnetic Kerr rotation [146] and also from the experiments by the neutron inelastic scattering[145], $t_{a_{1g}^* a_{1g}^*} = 0.2$, $t_{b_{1g} b_{1g}} = 0.4$, $t_{a_{1g}^* b_{1g}} = \sqrt{t_{a_{1g}^* a_{1g}^*} t_{b_{1g} b_{1g}}} \sim 0.28$, $\varepsilon_{a_{1g}^*} = 0$, $\varepsilon_{b_{1g}} = 2.6$, $K_{a_{1g}^*} = -2.0$, $K_{b_{1g}} = 4.0$ in units of eV, where the values of Hund's coupling exchange constant $K_{a_{1g}^*}$ and Zhang–Rice exchange constant $K_{b_{1g}}$ are taken from the first principles cluster calculations for a CuO$_6$ octahedron in LSCO [104], and the energy difference of the effective one-electron energies between a$_{1g}^*$ and b$_{1g}$ orbital states, $\varepsilon_{b_{1g}} - \varepsilon_{a_{1g}^*} = 2.6$, is determined so as to reproduce the energy difference between the ^{3}B$_{1g}$ and ^{1}A$_{1g}$ multiplets in the MCSCF-CI cluster calculations [103]. On the other hand, the values of

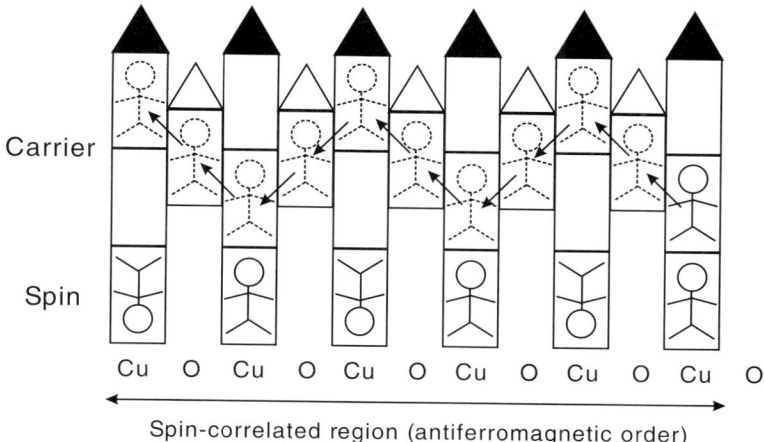

Fig. 8.4. Illustration of the two story house model and of how a hole carrier with up-spin itinerates within a spin-correlated region. The *lower story* consists of the Cu localized spins, which form the antiferromagnetic ordering in a spin-correlated region. In the *upper story*, a carrier with up-spin enters into the Cu-second-floor due to the Hund's coupling effect with Cu localized up-spins in the *lower story* (Hund's coupling triplet) while it enters into the hybridized state of O-bridge and Cu-third-floor at the Cu houses with localized down-spins due to the antiferromagnetic interaction between the carrier's and localized spins (Zhang–Rice singlet)

$t_{a_{1g}^* a_{1g}^*}$ and $t_{b_{1g} b_{1g}}$ are chosen so as to reproduce the antibonding a_{1g}^* and bonding b_{1g} bands in the band structure of La$_2$CuO$_4$ [106, 147, 148]. When we solve the K–S Hamiltonian for the multiplets of the Hund's coupling triplet and of the Zhang–Rice singlet in a CuO$_6$ octahedron by using the parameters determined above, we find that energy difference between the highest occupied levels in the Hund's coupling triplet and Zhang–Rice singlet which include the effect of the exchange interaction H_{ex} is 0.1 eV. Thus the highest levels of the Hund's coupling triplet and Zhang–Rice singlet are mixed by transfer interaction of $t_{a_{1g}^* b_{1g}}$ (~ 0.28).

As a result a hole-carrier moves coherently by taking the $^3B_{1g}$ to $^1A_{1g}$ multiplets alternately in the presence of the local AF order. Thus the K–S model holds.

8.4 Concluding Remarks

Originally the Kamimura–Suwa model was named by one of the present authors (Kamimura) as *"two-story house model"* [27], before HTSC researchers called it the K–S model. That is, looking at Figs. 8.1, 8.2 and 8.3, Kamimura called the K–S model and the K–S Hamiltonian in (8.3) the "two story house model". According to him, the upper story corresponds to the carrier states consisting of two kind of orbitals a_{1g}^* and b_{1g}, while the lower story corresponds to the system of the localized spins, as shown in Fig. 8.4. Each house represents a Cu house, in which the lower story corresponds to the b_{1g}^* orbital while the upper story consists of the second and third floors with characters of a_{1g}^* and b_{1g} orbitals, respectively. In the lower story a Cu d-hole with $d_{x^2-y^2}$ character is localized with up spin or down spin. They form an AF order by the superexchange interaction in the spin correlated region. As for the upper story, the room is connected with the neighbouring Cu houses by bridges of oxygen p_σ orbitals (the b_{1g} orbitals). As a result a hole carrier in the upper story move from an a_{1g}^* second floor room in a Cu house with an up-localized spin (right-hand side of the figure) to a b_{1g} third floor room in the neighbouring Cu house with down-localized spin through the O p_σ bridge, etc. Thus a metallic state is created in the upper story for an area of spin-correlated region while the lower story contributes a local AF order in the spin-correlated region.

9 Exact Diagonalization Method to Solve the K–S Hamiltonian

9.1 Introduction

In the previous chapter we described the essence of the K–S model. A problem with which we are now going to be concerned is how to solve the K–S Hamiltonian H_{KS} in (8.3), which was described in detail in the Sect. 8.3. In this chapter we try to solve the K–S Hamiltonian exactly for a system of a finite size. Simultaneously we investigate the validity of the K–S model when a single hole-carrier is doped into 1D chain and 2D square quantum spin systems, and also when two hole-carriers are doped.

For the purpose of investigating the validity of the K–S model, we first calculate the spin-correlation function and the orbital correlation function, by diagonalizing the K–S Hamiltonian with the exact diagonalization method in the cases of a single hole-carrier and two hole-carriers. The calculated results for the K–S model are presented in Sects. 9.3, 9.4 and 9.5. In order to clarify an important role of the coexistence of two kinds of orbitals, a_{1g}^* and b_{1g} orbitals in the K–S model mentioned in Chap. 8, we investigate the case where there is only a single orbital state by taking the energy difference between two kinds of orbitals, $\Delta \epsilon \, (= \varepsilon_{b_{1g}} - \varepsilon_{a_{1g}^*})$, in the K–S Hamiltonian to be infinity. This corresponds to the case of the Zhang–Rice singlet.

Further we calculate the radial distribution function between two hole-carriers, which is a characteristic quantity for a two-carrier system. The calculated results are presented in Sect. 9.6.

9.2 Description of the Method: Lanczos Method

In this chapter we adopt the exact diagonalization method. In order to solve the K–S Hamiltonian as accurately as possible, we have to adopt a model system of finite size. As an object of study we consider a two-dimensional square lattice with 4×4 sites shown in Figs. 9.1(a) and (b), because in cuprates a CuO_2 plane which is similar to a 2D square lattice plays an important role. We also consider a 1D chain lattice with L ($4 \leq L \leq 20$) sites in order to investigate the effect of dimensionality.

The representations of the site numbers in the 2D square lattice and in the 1D chain lattice are shown in Figs. 9.1(a) and (c), respectively. In the 2D

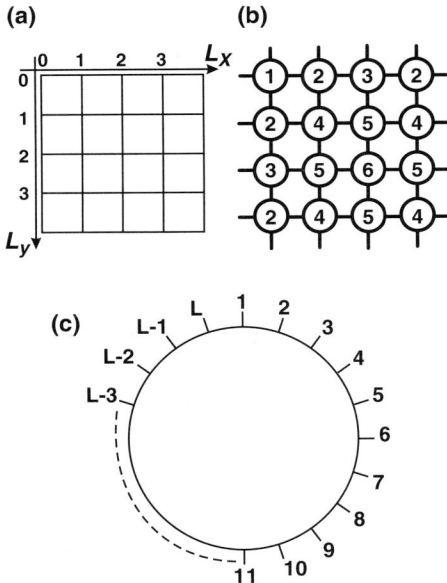

Fig. 9.1. Schematic views of the model systems: (**a**) 2D square lattice; (**b**) the numbered sites in the 2D square lattice, where equivalent sites are denoted by the same number. (**c**) 1D chain lattice with numbering sites

square system the equivalent sites are chosen as shown in Fig. 9.1(b). The $S = 1/2$ spins which are placed at each site represent the localized spins in the K–S model. They are coupled by the superexchange interactions $J\boldsymbol{S}_i \cdot \boldsymbol{S}_j$, the third term in (8.3), to form the AF order in the absence of dopant carriers. Applying the Lanczos method to these systems with the periodic boundary condition (PBC), we study the ground state for the K–S Hamiltonian in the following two cases; (1) a single doped hole-carrier ($N = 1$) and (2) two doped hole-carriers ($N = 2$).

In order to obtain the ground state based on the the K–S model, we have to diagonalize the K–S Hamiltonian $H_{\rm KS}$ for a system of finite size. If the size of a system is small enough to be diagonalized, all the matrix elements of the K–S Hamiltonian, $H_{\rm KS}$, are stored within the computational memories at one time. However, if a system size exceeds around 10 sites for the K–S model, it is difficult to diagonalize $H_{\rm KS}$ exactly due to the memory limitations. The K–S model with a single dopant hole has 8 states per site, because there appear two multiplets, $^1A_{1g}$ and $^3B_{1g}$, for either of the up and down localized spins in the AF order. Thus a finite system of L sites has 8^L states in principle, which is $\sim 1.0 \times 10^9$ for $L = 10$. In this context we pay attention to the ground state and the low-lying excited states. In this case,

9.2 Description of the Method: Lanczos Method

among the exact diagonalization methods, the Lanczos method [16, 149, 150] is appropriate. Let us describe the algorithm of the Lanczos method below.

Suppose that the Hamiltonian H which we want to solve is transformed to tridiagonal H'. When a transformation which transforms to tridiagonal H' is denoted by P, we have the relation of $P^{-1}HP = H'$. First the initial vector $|u_0\rangle$ in the Hilbert space of the model being study is selected. Then we generate a new vector by applying the Hamiltonian H to the initial vector $|u_0\rangle$. Consequently we obtain

$$|u_1\rangle = H|u_0\rangle - \frac{\langle u_0|H|u_0\rangle}{\langle u_0|u_0\rangle}|u_0\rangle , \tag{9.1}$$

where $\langle u_0|u_1\rangle = 0$ holds. Here $\langle u|$ represents a transposed vector of a vector $|u\rangle$. Further we are able to obtain a new vector $|u_2\rangle$ that is orthogonal to the previous two vectors $|u_0\rangle$ and $|u_1\rangle$ as

$$|u_2\rangle = H|u_1\rangle - \frac{\langle u_1|H|u_1\rangle}{\langle u_1|u_1\rangle}|u_1\rangle - \frac{\langle u_1|u_1\rangle}{\langle u_0|u_0\rangle}|u_0\rangle . \tag{9.2}$$

It can be easily checked that $\langle u_1|u_2\rangle = \langle u_0|u_2\rangle = 0$. This procedure can be generalized by defining the nth orthogonal basis recursively as

$$|u_{n+1}\rangle = H|u_n\rangle - \alpha_n|u_n\rangle - \beta_{n-1}|u_{n-1}\rangle , \tag{9.3}$$

where $n = 0, 1, 2, \ldots$, and the coefficients α_n and β_n are given by

$$\alpha_n = \frac{\langle u_n|H|u_n\rangle}{\langle u_n|u_n\rangle} , \tag{9.4}$$

$$\beta_{n-1} = \frac{\langle u_n|u_n\rangle}{\langle u_{n-1}|u_{n-1}\rangle} . \tag{9.5}$$

In this basis the tridiagonal Hamiltonian matrix H' can be expressed as

$$H' = \begin{pmatrix} \alpha_0 & \beta_1 & 0 & 0 & \ldots \\ \beta_1 & \alpha_1 & \beta_2 & 0 & \ldots \\ 0 & \beta_2 & \alpha_2 & \beta_3 & \ldots \\ 0 & 0 & \beta_3 & \alpha_3 & \ldots \\ \vdots & \vdots & \vdots & \vdots & \ddots \end{pmatrix} . \tag{9.6}$$

In this way the tridiagonal H' is diagonalized easily with use of ordinary library subroutines such as the bisection method, etc. However, it is impossible to completely diagonalize the K–S Hamiltonian, H_{KS}, because a number of interactions equal to the size of Hilbert space are needed. In practice, this would demand a considerable amount of CPU time. However, one of the advantages of this method is that adequately accurate information about the ground state of H_{KS} can be obtained after a small number of iterations. In the present case, we are able to diagonalize H' by a number of iterations of the order of ~ 100.

In the present calculations the z-component of the total spins in the 2D and 1D systems, S^z_{total}, is fixed to be a minimum value. For example, in the case of a 2D system of 16 sites with a single dopant hole, the minimum value of the z-component of the total spins is $1/2$ ($S^z_{\text{total}} = 1/2$) since the total number of the spins of $S = 1/2$ in the AF order and in the hole-carrier system is 17. In the case of a 2D system of 16 sites with two dopant holes, on the other hand, the minimum value of the total spins is 0 since the number of spins in this case is 18. All the bases which construct a wavefunction in the above two cases satisfy a minimum value of the z-component of the total spins. Besides, in the case of two dopant holes we have constructed a wavefunction so as to satisfy the Pauli principle.

As regards the parameters in (8.3), we adopt the values described in Chap. 8. These are $J = 0.1$, $K_{a^*_{1g}} = -2.0$, $K_{b_{1g}} = 4.0$, $t_{a^*_{1g} a^*_{1g}} = 0.2$, $t_{b_{1g} b_{1g}} = 0.4$, $t_{a^*_{1g} b_{1g}} = \sqrt{t_{a^*_{1g} a^*_{1g}} t_{b_{1g} b_{1g}}} \sim 0.28$, $\varepsilon_{a^*_{1g}} = 0$, $\varepsilon_{b_{1g}} = 2.6$ in units of eV.

By using bases mentioned above and these parameters, we solve the effective Hamiltonian of the K–S model in (8.3).

9.3 Calculated Results for the Spin-Correlation Functions

In discussing the calculated results, it is helpful to consider first a case without spin fluctuations in the localized spin system in order to appreciate the nature of the electronic state of a single dopant hole. For this purpose, let us suppose that the complete antiferromagnetic (AF) ordering, i.e., Néel order, has been established among the localized spins. In this case we have only to consider a term of z-component $S^z_i S^z_j$ and $s^z_{i,m} S^z_i$ in the Heisenberg Hamiltonian H_{AF} and H_{ex} in (8.3). Then the ground-state energy for the 2D square lattice of L sites with a single dopant hole is obtained, in units of eV, as

$$E^{1h}_{\text{Neel}} = -\frac{1}{4} zJL \times \frac{1}{2} - 1.517 , \qquad (9.7)$$

where L is the number of the localized spins, and z is the number of nearest bond around a site. We obtained $E^{1h}_{\text{Neel}} = -2.317$ eV in (9.7) with $z = 4$ and $L = 16$. In this case the wavefunction of a hole is extended over the whole system and the numerical values for the squares of the components a^*_{1g} and b_{1g} orbital states in the wavefunction for a dopant hole with up-spin are shown in Fig. 9.2. Although there is a tendency that a dopant hole alternately occupies the two orbitals, a^*_{1g} and b_{1g}, site by site [15] (a "zigzag" like state), its wavefunction spills out due to a quantum-mechanical tunneling effect, so that every orbital component has a finite amplitude at each site.

In order to fully take into account the spin fluctuation effect of the AF order due to the localized hole spins, Hamada, Ishida, Kamimura, and Suwa

9.3 Calculated Results for the Spin-Correlation Functions

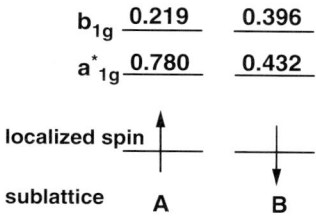

Fig. 9.2. Numerical values for the squares of a_{1g}^* and b_{1g} orbital states in the wavefunction of the hole with up-spin in the case where the spins of localized holes form the AF order in the Néel state. The direction of the localized spins at A and B sublattice is upwards and downwards, respectively

[151] carried out the exact diagonalization of the K–S Hamiltonian using the Lanczos method for a 2D square lattice system with 16 (4×4) sites which consists of 16 localized spins and a single itinerant hole, applying the periodic boundary condition. We note that, for a 4 × 4 square lattice with a periodic boundary condition, there are only six sites which are not equivalent. These independent sites are numbered as shown in Fig. 9.1(b). The localized spins at these sites are classified as A and B by two sublattices, depending on the directions of up or down spins when there are no dopant holes. The calculated ground-state energy of (8.3) with a single hole is $E_g^{1h} = -2.934 \,\mathrm{eV}$. This is lower than $E_{\mathrm{Neel}}^{1h} (= -2.317 \,\mathrm{eV})$ in (9.7) with $z = 4$ and $L = 16$. Hereafter we omit the unit of energy eV. The difference of energy ΔE_g^{1h}, which is defined as $\Delta E_g^{1h} = E_g^{1h} - E_{\mathrm{Neel}}^{1h}$, is -0.617. This means that the ground state in the K–S Hamiltonian is more stable due to the spin fluctuation effect, compared with that in the Néel order system.

On the other hand, we consider the case of a spin system without any dopant hole (un-doped case). It is well known that the energy of ground state for the antiferromagnetic Ising model is represented as $-\frac{1}{4}JzL \times \frac{1}{2}$. This means that the ground state of the antiferromagnetic Ising spin system is consistent with the Néel state. When the ground-state energy in the un-doped case is denoted by E_{Neel}^{0h}, the calculated value of E_{Neel}^{0h} is -0.8 for a 2D square lattice with 16 sites. If the spin fluctuation is introduced into the antiferromagnetic Ising spin system, the antiferromagnetic Heisenberg model is obtained where the ground-state energy for the latter is calculated to be $E_g^{0h} = -1.123$. The energy difference ΔE_g^{0h}, defined by $\Delta E_g^{0h} = E_g^{0h} - E_{\mathrm{Neel}}^{0h}$, is -0.323. Further Hamada et al. [151] calculated the energy difference between E_g^{1h} and E_g^{0h}, ΔE_g^{1-0}, and that between E_{Neel}^{1h} and E_{Neel}^{0h}, $\Delta E_{\mathrm{Neel}}^{1-0}$. These are $\Delta E_g^{1-0} = -1.811$ and $\Delta E_{\mathrm{Neel}}^{1-0} = -1.517$. Thus $\Delta \equiv \Delta E_g^{1-0} - \Delta E_{\mathrm{Neel}}^{1-0} = -0.294$. From this fact and from the comparison between ΔE_g^{1h} and ΔE_g^{0h}, we can say that the K–S model with a single dopant hole is more stabilized not only by the spin fluctuation effect but also by the exchange

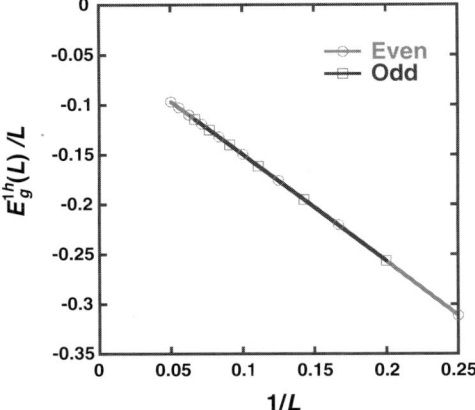

Fig. 9.3. The dependence of the ground-state energy per lattice-site L, $E_g^{1h}(L)/L$, on the lattice-size L for a 1D chain lattice of L sites with a single dopant carrier. The *circles* represent the ground-state energy per lattice-site in the case that L is an even number. The *squares* correspond to the case when L takes an odd number

interaction between the spins of a dopant hole and a localized hole, H_{ex} in (8.3).

Hamada et al. [151] also calculated the ground-state energy in a 1D chain lattice with L sites. It is impossible to calculate when varying the size of lattice in a 2D system due to the memory limitations. In a 1D chain system, however, we are able to solve the K–S Hamiltonian within the range of 4 sites to 20 sites. Let us denote the ground-state energy for L sites with a single dopant carrier by $E_g^{1h}(L)$. The calculated result of $E_g^{1h}(L)$ is shown in Fig. 9.3. In the case of the 1D Heisenberg model (un-doped case), generally there are two curves for the size dependence of the ground-state energy corresponding to the following two cases: (1) The lattice site number L is an odd number; (2) that of L is an even number. The separation between two curves is large due to the effect of the spin frustration effect in the AF order in the localized spin system when a size of the system is small. On the other hand, in the case of the K–S model in a 1D system, the system-size dependence of the ground-state energy is quite different. There is no separation between two curves corresponding to two cases where L is an odd number and an even number, even though the system size is small. In fact, there is only a single line in a relation of energy vs. system size, as shown in Fig. 9.3. This means that a dopant hole which itinerates in a system can suppress the energy loss due to the spin frustration in the AF order for the K–S model.

In this section, we investigate how much the AF order survives in the localized spin system in the presence of dopant holes, by calculating the *spin-correlation function* between the localized spins. When the number of dopant holes in the system is $N(\geq 1)$, the *spin-correlation function* between

9.3 Calculated Results for the Spin-Correlation Functions

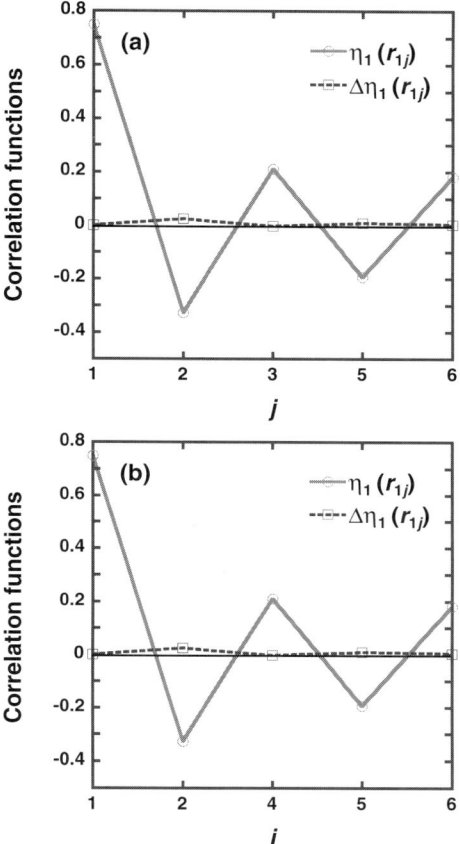

Fig. 9.4. The calculated results of the spin-correlation function $\eta_1(\boldsymbol{r}_{1j})$ and the difference of spin-correlation function from that of the un-doped system $\Delta\eta_1(\boldsymbol{r}_{1j})$ along two paths of (**a**) 1-2-3-5-6 and (**b**) 1-2-4-5-6 in a 2D square lattice, as a function of the distance between sites 1 and j

the localized spins is defined by

$$\eta_N(\boldsymbol{r}_{ij}) = \langle \boldsymbol{S}_i \cdot \boldsymbol{S}_j \rangle , \tag{9.8}$$

where $\boldsymbol{r}_{ij}$ is a position vector connecting sites i and j, where $\boldsymbol{r}_{ij} = \boldsymbol{r}_i - \boldsymbol{r}_j$. The calculated results of the spin-correlation function for the 2D and the 1D systems with a single hole-carrier, $\eta_1(\boldsymbol{r}_{1j})$, are shown in Figs. 9.4(a) and (b) and Fig. 9.5, respectively. These functions are as a function of the distance between sites j and 1.

Then we compare the values of $\eta_1(\boldsymbol{r}_{1j})$ with the spin-correlation function calculated for a 2D AF system with no dopant holes (un-doped case), which is denoted by $\eta_u(\boldsymbol{r}_{1j})$. The difference between $\eta_1(\boldsymbol{r}_{1j})$ and $\eta_u(\boldsymbol{r}_{1j})$, $\Delta\eta_1(\boldsymbol{r}_{1j})$

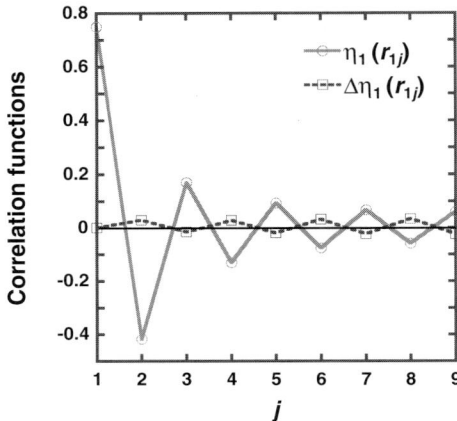

Fig. 9.5. The calculated results of the spin-correlation function $\eta_1(\mathbf{r}_{1j})$ and the difference of spin-correlation function from that of the un-doped system $\Delta\eta_1(\mathbf{r}_{1j})$ in a 1D chain lattice of 16 sites, as a function of the distance between sites 1 and j. Due to the periodic boundary condition, a horizontal axis is taken to $j = L/2 + 1$. As a result the results from $j = 1$ to 9 is shown

($= \eta_1(\mathbf{r}_{1j}) - \eta_u(\mathbf{r}_{1j})$), is shown by dotted lines in Fig. 9.4 for the case of $i = 1$. We note in Fig. 9.1(b) that there are two paths connecting sites 1 and 6, i.e., 1-2-3-5-6 and 1-2-4-5-6. These paths show that the sites on the two sublattices appear alternately. The calculated results for the correlation functions corresponding to the paths 1-2-3-5-6 and 1-2-4-5-6 are shown in Fig. 9.4(a) and Fig. 9.4(b), respectively. From these results, it is concluded that the AF ordering in the localized spin system is not destroyed even in the presence of a dopant hole. In fact, $|\Delta\eta_1(\mathbf{r}_{1j})|$ is considerably smaller than $|\eta_u(\mathbf{r}_{1j})|$ in the hole-doped system. In other words, this result means that a dopant hole in a 4×4 2D lattice does not have a significant effect on the AF order.

On other hand, the calculated results for the spin-correlation functions $\eta_1(\mathbf{r}_{1j})$ and the difference of the spin-correlation functions $\Delta\eta_1(\mathbf{r}_{1j})$ in a 1D system are shown in Fig. 9.5. It is seen from Fig. 9.5 that $|\Delta\eta_1(\mathbf{r}_{1j})|$ in a 1D system approaches to $|\eta_u(\mathbf{r}_{1j})|$ as j becomes larger. Thus we can say that the a dopant hole considerably affects a whole localized spin system in the case of a 1D system.

From the above calculated results, we conclude that the AF order in the localized spin system in the 2D system is not influenced by doping a single hole-carrier while that in the 1D system is disturbed, as far as the K–S model is concerned.

Now we proceed to investigate whether the AF order is influenced by doping two hole-carriers for 1D and 2D systems, based on the K–S model. For this purpose we calculate the spin-correlation function $\eta_2(\mathbf{r}_{ij})$ between

9.3 Calculated Results for the Spin-Correlation Functions

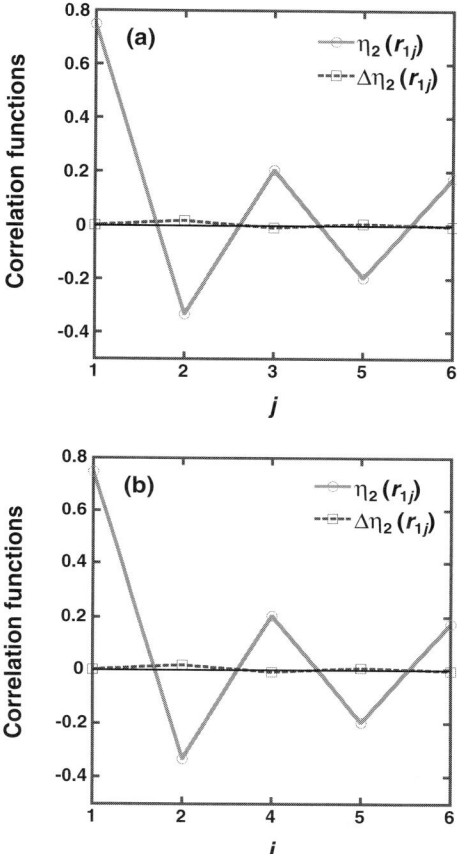

Fig. 9.6. The calculated results of the spin-correlation function $\eta_2(r_{1j})$ and its difference from that of the un-doped system $\Delta\eta_2(r_{1j})$ in a 2D square lattice of 4×4 sites with two hole-carriers (**a**) for the path 1-2-3-5-6 and (**b**) for the path 1-2-4-5-6. These quantities are shown as a function of the distance between sites 1 and j

the localized spins at sites $\boldsymbol{r}_i$ and $\boldsymbol{r}_j$ in the presence of *two* hole-carriers, and the difference of spin-correlation functions between an un-doped case and a two-hole case, $\Delta\eta_2(\boldsymbol{r}_{ij})$, which has been defined in (9.8), where $\boldsymbol{r}_{ij} = \boldsymbol{r}_i - \boldsymbol{r}_j$. Like the case of a single hole-carrier, we calculate these quantities for a 1D chain lattice and a 2D square lattice system.

First the calculated results of $\eta_2(\boldsymbol{r}_{1j})$ for a 2D square lattice are shown in Figs. 9.6(a) and (b), as a function of the distance between sites j and 1. Similarly to the case of a single hole-carrier, we show $\eta_2(\boldsymbol{r}_{1j})$ along the two-paths 1-2-3-5-6 and 1-2-4-5-6 in Figs. 9.6(a) and (b), respectively. Then we compare the calculated values of $\eta_2(\boldsymbol{r}_{1j})$ with the spin-correlation function

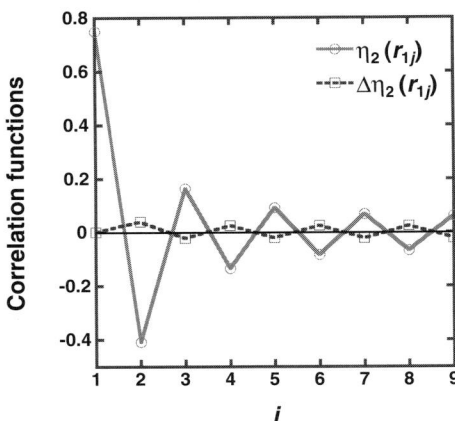

Fig. 9.7. The calculated results of the spin-correlation function $\eta_2(\boldsymbol{r}_{1j})$ and its difference from that of the un-doped system $\Delta\eta_2(\boldsymbol{r}_{1j})$ in a 1D chain lattice of 16 sites with two hole-carriers, where these quantities are shown as a function of the distance between sites 1 and j. Due to the periodic boundary condition, a horizontal axis is taken to be $j = L/2 + 1$. Thus it is enough to show the result from $j = 1$ to 9

for the 2D AF system with no holes (the un-doped case), the latter of which is denoted by $\eta_u(\boldsymbol{r}_{1j})$. The difference between $\eta_2(\boldsymbol{r}_{1j})$ and $\eta_u(\boldsymbol{r}_{1j})$, $\Delta\eta_2(\boldsymbol{r}_{1j})$ ($= \eta_2(\boldsymbol{r}_{1j}) - \eta_u(\boldsymbol{r}_{1j})$), is shown by dotted lines in Fig. 9.6. From these results, it is concluded that the 2D AF order in the localized spin system is not destroyed even in the presence of two dopant holes. This feature is similar to the one obtained in the case of a single dopant hole in the localized spin system.

Now the calculated results of $\eta_2(\boldsymbol{r}_{1j})$ for a 1D chain lattice are shown in Fig. 9.7, as a function of the distance between j and 1 along a chain. Comparing the results in Fig. 9.7 with those in the Fig. 9.6, $\Delta\eta_2(\boldsymbol{r}_{1j})$ for the 1D case is a little larger than $\Delta\eta_2(\boldsymbol{r}_{1j})$ for the 2D case. This means that the spin fluctuation effect in a 1D system is more remarkable than that in a 2D system, as is well known. As a result the K–S model is a little disturbed in a 1D system.

In summary, we have presented the calculated results of $\eta(\boldsymbol{r}_{1j})$ and $\Delta\eta(\boldsymbol{r}_{1j})$ for a 1D chain lattice and a 2D square lattice in the two cases of a single hole-carrier and two hole-carriers. In order to clarify an effect by hole-carriers on the AF order in the localized spin system, we have investigated the difference of the spin-correlation functions between an un-doped case and an N-carrier case with $N = 1$ or 2 for a 1D chain system and a 2D square system. By comparing the results in Fig. 9.8(a) with those in Fig. 9.8(b), we can say that the AF order in the localized spin system is more disturbed by the presence of carriers in the 1D chain system than in the 2D square system.

9.3 Calculated Results for the Spin-Correlation Functions

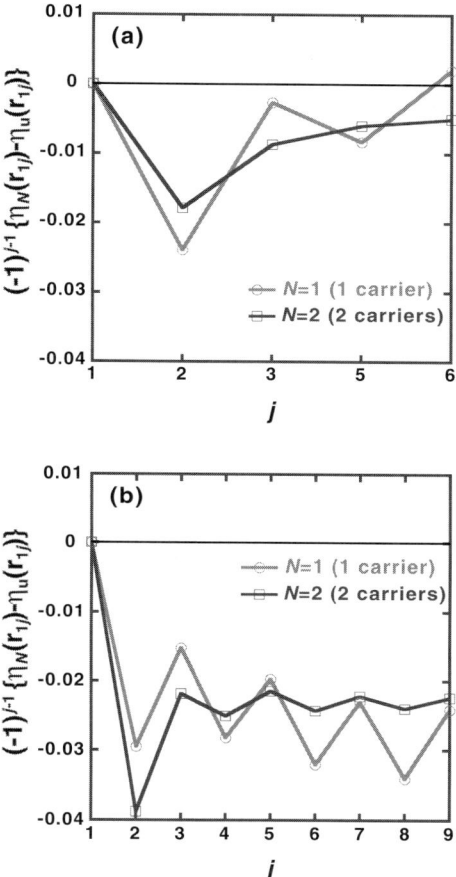

Fig. 9.8. The calculated results derived from the difference of the spin-correlation function from that of the un-doped system, $\Delta\eta_N(r_{1j})$, (**a**) in a 2D square system and (**b**) in a 1D chain system. A perpendicular axis represents the value of $(-1)^{j-1}\Delta\eta_N(r_{1j})$. Factor $(-1)^{j-1}$ is multiplied by $\Delta\eta_N(r_{1j})$ in order to clarify the change of the spin-correlation function by doping the hole-carriers

Further we can see that the disturbance in the AF order is more remarkable in a single carrier case than in the two-hole case. However, the AF order around $j=2$ in the 1D chain system is more disturbed in the two hole case while the AF order in the single hole case is disturbed over a long distance with zigzag behaviour. By comparing the results of the single hole case with two hole case, we may conclude that the K–S model holds more favorably when the carrier concentration increases.

9.4 Calculated Results for the Orbital Correlation Functions

Next we investigate whether a dopant hole is itinerant or not by examining a behaviour of its wavefunction in 1D and 2D systems. Although complete itinerancy can not be concluded from the calculated results of the present systems, it is nevertheless possible to discuss whether there is the tendency of the wavefunction of a dopant hole extending over the whole system. For this purpose, we calculated the *off-diagonal orbital correlation function* for a dopant hole defined by

$$\zeta_{mn}(\bm{r}_{ij}) = \sum_{\bm{r}'\sigma} \langle C^\dagger_{\bm{r}_{ij}+\bm{r}'m\sigma} C_{\bm{r}'n\sigma}\rangle \, , \qquad (9.9)$$

where $\zeta_{mn}(\bm{r}_{ij})$ represents the correlation between m-type orbital at the ith site and n-type orbital at jth site. If a dopant hole were localized, this correlation function would decay rapidly. We have calculated $\zeta_{mn}(\bm{r}_{ij})$ for the three cases; $m = n = \text{a}^*_{1\text{g}}$, $m = n = \text{b}_{1\text{g}}$, and $m = \text{a}^*_{1\text{g}}$ and $n = \text{b}_{1\text{g}}$, which we denote by $\zeta_{aa}(\bm{r}_{ij})$, $\zeta_{bb}(\bm{r}_{ij})$ and $\zeta_{ab}(\bm{r}_{ij})$, respectively.

First, we investigate the dependence of an off-diagonal orbital correlation function between the first nearest neighbour sites in a 2D system on the energy difference between $\text{a}^*_{1\text{g}}$ and $\text{b}_{1\text{g}}$ orbitals, $\Delta\epsilon$. If a dopant hole transfers like a metallic state based on the K–S model, the value of $\zeta_{ab}(\bm{r}_{ij})$ between the fist nearest neighbour sites should be at least larger than one of $\zeta_{aa}(\bm{r}_{ij})$ and $\zeta_{bb}(\bm{r}_{ij})$ between the first nearest neighbour sites. The calculated results of the $\Delta\epsilon$ dependence of off-diagonal orbital correlation function in a 2D system are shown in Fig. 9.9. From this figure we can see that, when the

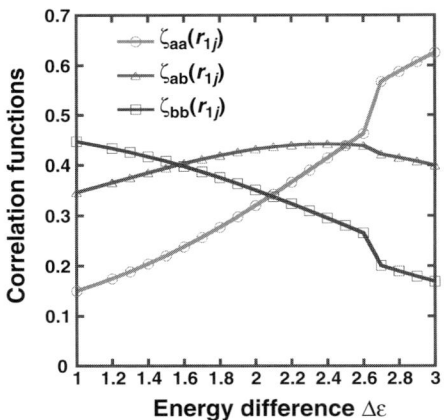

Fig. 9.9. The dependence of the off-diagonal orbital correlation functions between first nearest neighbour sites on the energy difference between $\text{a}^*_{1\text{g}}$ and $\text{b}_{1\text{g}}$ orbitals, $\Delta\epsilon$, for a 2D system with a single hole-carrier

9.4 Calculated Results for the Orbital Correlation Functions

energy difference between two orbitals takes a value from 1.5 eV to 2.6 eV, it is advantageous for a hole-carrier to move between two orbitals alternately.

Based on the results of the first-principles cluster calculation with the MCSCF-CI method for LSCO, Kamimura and Suwa obtained the value of the energy difference between a_{1g}^* and b_{1g} orbitals, $\Delta\epsilon$ in the K–S Hamiltonian, to be 2.6 eV. This value, derived from the results of the first-principles calculation, lies within the energy range from 1.5 eV to 2.6 eV suggested from Fig. 9.9. This result supports the validity of the K–S model. But it is insufficient to discuss the itinerancy of a dopant hole only from the off-diagonal orbital correlation function between the first nearest neighbour sites. Accordingly, for $\Delta\epsilon = 2.6$ eV in a 2D system with a single dopant hole, we investigate the off-diagonal orbital correlation function for a distant site from $j = 1$. Figures 9.10(a) and (b) show how $\zeta_{aa}(\boldsymbol{r}_{1j})$, $\zeta_{bb}(\boldsymbol{r}_{1j})$ and $\zeta_{ab}(\boldsymbol{r}_{1j})$ vary with site j along the paths 1-2-3-5-6 and 1-2-4-5-6, respectively, where site i is taken to be 1. As seen in Fig. 9.10, $\zeta_{mn}(\boldsymbol{r}_{1j})$ does not decay over the size of the system, suggesting that a hole does not localize even in the presence of AF order.

On the other hand, in the same procedure as the case of a 2D system, the dependence of the off-diagonal orbital correlation function between the first nearest neighbour sites for the case of a 1D system is shown in Fig. 9.11. As shown in Fig. 9.11, the off-diagonal orbital correlation functions in a 1D system are very sensitive to the change of the energy difference between two orbitals, $\Delta\epsilon$. In contrast to the case of a 2D square lattice, the energy region in which a dopant hole can hop between two orbitals is very narrow. We can see from Fig. 9.11 that when $\Delta\epsilon$ takes a value in the energy range from 2.3 eV to 2.6 eV, the itinerant motion of a dopant hole becomes possible. Like the case of a 2D system, the calculated results of the off-diagonal orbital correlation function at a site far from a first nearest neighbour site in a 1D chain lattice with a single dopant hole is shown in Fig. 9.12. As j increases, $\zeta_{aa}(\boldsymbol{r}_{1j})$, $\zeta_{ab}(\boldsymbol{r}_{1j})$ and $\zeta_{bb}(\boldsymbol{r}_{1j})$ decrease gradually. However, these values are so small that a metallic state based on the K–S model is not possible in a 1D system.

According to the present results for the Heisenberg spin Hamiltonian for the localized spin system in (8.3), $\zeta_{aa}(\boldsymbol{r}_{1j})$, $\zeta_{bb}(\boldsymbol{r}_{1j})$ and $\zeta_{ab}(\boldsymbol{r}_{1j})$ show an oscillating behaviour with changing site j. From this result we can say that the spin fluctuation in the AF order assists the "zigzag"-like states. To be more precise, an alternating behaviour of the increase and decrease in the magnitudes of $\zeta_{aa}(\boldsymbol{r}_{ij})$, $\zeta_{bb}(\boldsymbol{r}_{ij})$ and $\zeta_{ab}(\boldsymbol{r}_{ij})$ appears when $\boldsymbol{r}_{ij}$ connecting two sites varies between different sublattices. This behaviour is due to the aid of the spin fluctuation effect in the localized spin system.

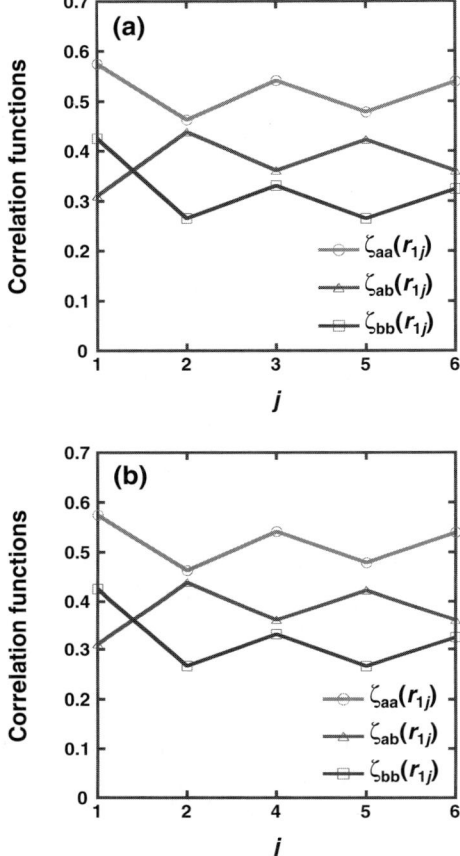

Fig. 9.10. The calculated results of the off-diagonal orbital correlation functions for a dopant hole; $\zeta_{aa}(\mathbf{r}_{1j})$, $\zeta_{ab}(\mathbf{r}_{1j})$ and $\zeta_{bb}(\mathbf{r}_{1j})$: (**a**) for the path 1-2-3-5-6 and (**b**) for the path 1-2-4-5-6. In these calculations the energy difference between a_{1g}^* and b_{1g} orbitals is chosen to be 2.6 eV

9.5 The Case of a Single Orbital State

In this section, we investigate the case where only a single orbital state exists in the Hamiltonian (8.3) in order to understand the importance of the coexistence of the two kinds of orbital states a_{1g}^* and b_{1g} in the metallic state in cuprates, especially in a 2D system. In particular, we pay a special attention to the case where only the Zhang–Rice singlet state exists [96]. In order to treat this special case the absolute value of $\varepsilon_{b_{1g}}$ ($\varepsilon_{b_{1g}} < 0$) in the K–S Hamiltonian (8.3) is taken to be very large, compared with $\varepsilon_{a_{1g}^*}$. As a result, only

9.5 The Case of a Single Orbital State 77

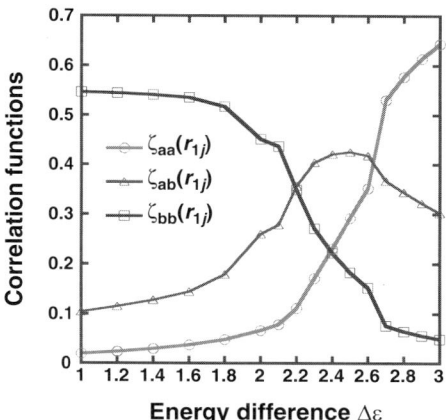

Fig. 9.11. The dependence of the off-diagonal orbital correlation functions between first nearest neighbour sites on the energy difference between a_{1g}^* and b_{1g} orbitals, $\Delta\epsilon$, for a 1D chain system with a single hole-carrier

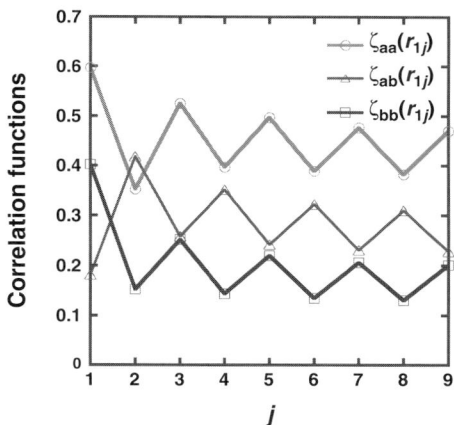

Fig. 9.12. The calculated results of the off-diagonal orbital correlation functions in a 1D chain system with a dopant hole; $\zeta_{aa}(r_{1j})$, $\zeta_{ab}(r_{1j})$ and $\zeta_{bb}(r_{1j})$, when the energy difference between a_{1g}^* and b_{1g} orbitals is 2.6 eV

the b_{1g} orbital state plays a role. The remaining parameters are taken as the same as those in the case where the two orbitals a_{1g}^* and b_{1g} coexist.

We call this case *"one-orbital case"*. Apparently the ground state in this case consists of the Zhang–Rice singlets, where a dopant hole occupies only the b_{1g} orbital. Figure 9.13 shows the calculated spin-correlation function ($\eta_1(r_{1j})$) and its difference from $\eta_u(r_{1j})$, i.e., $\Delta\eta_1(r_{1j})$ for the "one-orbital case". This result clearly shows that $\Delta\eta_1(r_{1j})$ is comparable to $\eta_1(r_{1j})$ and thus in the "one-orbital case" the AF order is destroyed. As for the

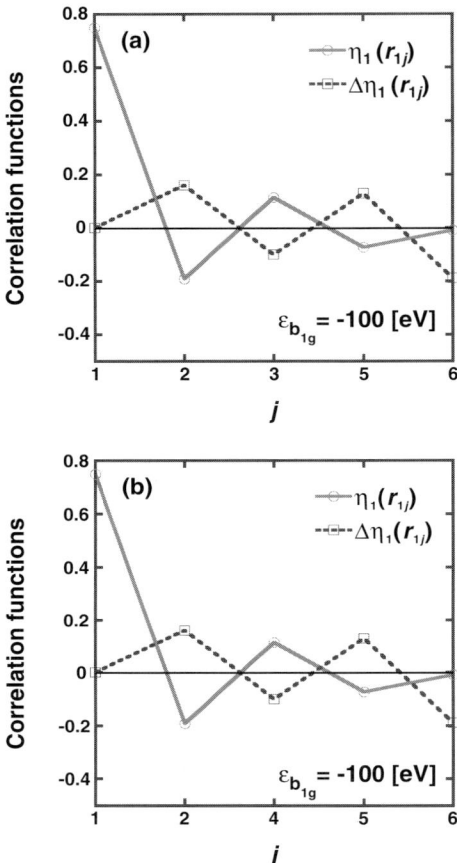

Fig. 9.13. The calculated results of the spin-correlation function $\eta_1(\mathbf{r}_{1j})$ and the difference of spin-correlation function from that of the un-doped system $\Delta\eta_1(\mathbf{r}_{1j})$ for $\varepsilon_{b_{1g}} = -100\,\mathrm{eV}$: (**a**) for the path 1-2-3-5-6 and (**b**) for the path 1-2-4-5-6

off-diagonal orbital correlation for a dopant hole, only $\zeta_{bb}(\mathbf{r}_{1j})$ is considered. The calculated result of $\zeta_{bb}(\mathbf{r}_{1j})$ is shown in Fig. 9.14 as a function of the distance between sites 1 and j. From this result we can say that a dopant hole is localized within the space of several sites. This may correspond to a situation similar to the case in which a localized spin-polaron is formed in a single band system. Here, in order to check this inference, we calculated the spin-correlation function between a spin of a dopant hole and a localized spin in the underlying AF lattice. The calculated results are shown in Fig. 9.15. From the results in Fig. 9.15 we may say that a spin polaron is formed around $j = 1$. Consequently we conclude that the coexistence of two orbitals a_{1g}^* and b_{1g} plays an important role in making a dopant hole itinerant in a cuprates

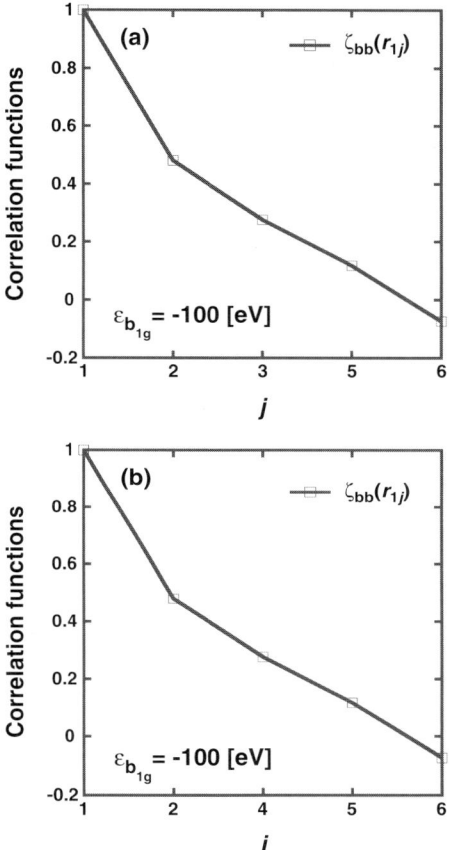

Fig. 9.14. The calculated results of the off-diagonal orbital correlation function for the dopant hole in the "Zhang–Rice case", $\zeta_{bb}(r_{1j})$, where we take $\varepsilon_{b_{1g}} = -100$ eV: (**a**) for the path 1-2-3-5-6 and (**b**) for the path 1-2-4-5-6

in the presence of AF order. Finally a remark is made on the t–J model [96]. In the t–J model the Zhang–Rice singlet is considered to be a quasi-particle while in the present calculation the Zhang–Rice singlet has been treated as one of the states in the ground state.

9.6 Calculated Results of the Radial Distribution Function for Two Hole-Carriers

The radial distribution function between two hole-carriers is introduced in order to investigate a spatial correlation between two hole-carriers. The radial

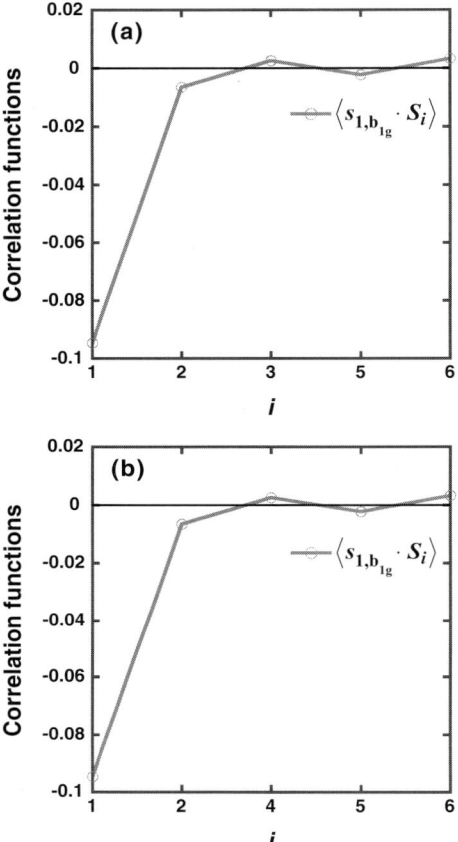

Fig. 9.15. The calculated results of the spin-correlation function between a spin of a hole-carrier in the b_{1g} orbital and localized spins in 2D AF order, $\langle s_{1,b_{1g}} \cdot S_i \rangle$, for $\varepsilon_{b_{1g}} = -100\,\text{eV}$. An ith site is chosen along two paths of (**a**) 1-2-3-5-6 and (**b**) 1-2-4-5-6, respectively

distribution function is denoted as $P_{mn}(R)$ for the distance between two hole-carriers, $R = |\boldsymbol{r}_i - \boldsymbol{r}_j|$. $P_{mn}(R)$ represents a spatial distribution of two hole-carriers for the case where one hole-carrier is located at a place $\boldsymbol{r}_i$ occupying the m-type orbital and the other is located at a place $\boldsymbol{r}_j$ occupying the n-type orbital. By using the number operator $\hat{n}_{i,m}(= \hat{n}_{i,m,\uparrow} + \hat{n}_{i,m,\downarrow})$, $P_{mn}(R)$ is defined as

$$P_{mn}(R) = \sum_{\langle i,j \rangle} \langle \hat{n}_{i,m} \hat{n}_{j,n} \rangle . \tag{9.10}$$

Here $\hat{n}_{i,m,\uparrow}$ and $\hat{n}_{i,m,\downarrow}$ are the number operators for the holes in m-type orbital at ith site with up-spin and down-spin, respectively. Further by taking

9.6 Calculated Results of the Radial Distribution Function   81

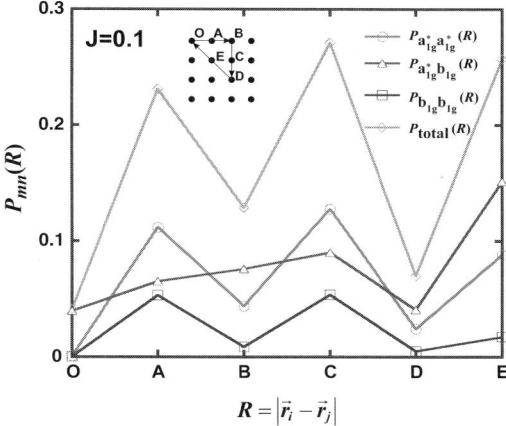

Fig. 9.16. The calculated results of the distribution functions of two hole-carriers; $P_{a_{1g}^* a_{1g}^*}(R)$, $P_{a_{1g}^* b_{1g}}(R)$ and $P_{b_{1g} b_{1g}}(R)$. Initial point of R is placed at O as shown in the illustration of figure, where the first hole lies at O

summation over all the orbitals ($m, n = a_{1g}^*$ or b_{1g}), $P_{\text{total}}(R)$ is defined as

$$P_{\text{total}}(R) = \sum_{m,n} P_{mn}(R) , \qquad (9.11)$$

where $P_{\text{total}}(R)$ is normalized to 1, that is $\int P_{\text{total}}(R) \, dR = 1$. There are three types of $P_{mn}(R)$; $m = n = a_{1g}^*$, $m = a_{1g}^*$ and $n = b_{1g}$, and $m = n = b_{1g}$, which are denoted as $P_{a_{1g}^* a_{1g}^*}(R)$, $P_{a_{1g}^* b_{1g}}(R)$ and $P_{b_{1g} b_{1g}}(R)$, respectively.

The calculated radial distribution functions of the three types $P_{mn}(R)$ and of $P_{\text{total}}(R)$ are shown in Fig. 9.16 for various sites in the 2D square lattice, A, B, C, D and E, where the first hole is located at O site and the second hole moves from O site to A, B, C, D and E. From Fig. 9.16 we can see that $P_{mn}(R)$ shows the zigzag-like behaviour. Further the following conclusions emerge from the calculated results shown in Fig. 9.16, in particular from the result for $P_{\text{total}}(R)$: (1) The case where two holes are located at O and C sites is the highest probability and the case where two holes are located at O and A sites appears with higher probability. This indicates the coulomb repulsion between hole-carriers is not strong. In both cases the directions of the localized spins in the AF order are opposite. The present calculated result is in good agreement with the experimental results on the coherent length of Cooper pair, which is 10 to 15 Å, because the distance between O and A site is nearly 10 Å. (2) A reason why $P_{a_{1g}^* b_{1g}}(R = 0)$ has a finite value instead of zero value is due to the fact that the two holes with opposite spins may come to the same site even though the Hubbard-U interaction exists.

10 Mean-Field Approximation for the K–S Hamiltonian

10.1 Introduction

In Chap. 9, we described one of the methods to solve the K–S Hamiltonian, i.e., the exact diagonalization method. However, since this method is applicable only to a system of a finite size, we have to develop a new method to solve the K–S Hamiltonian for a real cuprate material. For this purpose, in this chapter we will develop a method of solving the K–S Hamiltonian in an approximate way, which is called a mean-field approximation.

Let us now explain the mean-field approximation for the K–S Hamiltonian. In order to solve the K–S Hamiltonian (8.3) in Chap. 8, we tried to separate a hole-carrier system and a system of the localized spins which occupy the upper Hubbard b_{1g}^* band and form the antiferromagnetic (AF) ordering due to the superexchange interaction between the localized spins, J in (8.3). For this aim we assume that a spin-correlated region is widely spread, and we treat the the exchange interaction between the spins of a dopant hole and a localized spin in (8.3), $\sum_{i,m} K_m \boldsymbol{s}_{im} \cdot \boldsymbol{S}_i$, in the mean field approximation by replacing the localized spins $\boldsymbol{S}_i$'s by its average value $\langle \boldsymbol{S}_i \rangle$. This method was developed by Kamimura and Ushio, and numerical calculations have been performed by Ushio and Kamimura. In this method the values of $\boldsymbol{S}_i$ at $T = 0$ K are taken as the average values of $\langle \boldsymbol{S}_i \rangle$, that is $+1$ for A-site and -1 for B-site. Thus the effect of the localized spins is dealt with like a molecular field acting on a dopant hole. This approximation is called "mean-field approximation".

In order to derive the effective one-electron-type Hamiltonian for the dopant holes, we determine the "molecular field" of the localized spins so as to reproduce the results of first-principles calculation for a CuO_6 octahedron by Kamimura and Eto [104]. In other words, we determine the effective-one-electron type Hamiltonian in a periodic system so that the energy of $^3B_{1g}$ and $^1A_{1g}$ multiplets calculated by using the effective one-electron-type Hamiltonian coincides with that of first-principles cluster calculations by Kamimura and Eto, and further assume that the lifetime broadening effect due to the finite spin correlation length is neglected.

In this context the calculation of the effective one-electron-type band structure of the carrier system is performed by renormalizing the effects of the exchange integral between the spins of a dopant hole and a localized spin into the carrier states. In doing so one should first note that the holes which are accommodated in the antibonding b_{1g}^* orbitals are localized at Cu site by the strong U effect and the spins of localized holes in b_{1g}^* orbitals are coupled antiferromagnetically due to the superexchange interaction between the localized spins, the third terms with J in the right hand side of (8.3).

Since the dopant holes move coherently over a long distance without destroying the AF order, occupying from the high-spin $^3B_{1g}$ multiplet to the low-spin $^1A_{1g}$ multiplet and then to the high-spin $^3B_{1g}$ multiplet in the "molecular field" of the localized spins, we take a unit cell so as to contain two neighbouring CuO_6 octahedrons with up- and down-localized spins called A- and B-sites. Further, in order to realize the alternate appearance of b_{1g} and a_{1g}^* orbitals through O p_σ orbitals, we take into account the CuO_2 network structure explicitly and consider the 34 × 34 dimensional matrix ($\tilde{H}(\boldsymbol{k})$), where $2p_x$, $2p_y$ and $2p_z$ atomic orbitals for each of eight oxygen atoms and $3d_{yz}$, $3d_{xz}$, $3d_{xy}$, $3d_{x^2-y^2}$ and $3d_{z^2}$ atomic orbitals for each of two Cu atoms in the unit cell are taken as the basis functions. This Hamiltonian matrix $\tilde{H}(\boldsymbol{k})$ consists of two parts; the one-electron part $\tilde{H}_0(\boldsymbol{k})$, and the effective-interaction part $\tilde{H}_{\text{int}}(\boldsymbol{k})$, which comprises the many-body interactions such as the exchange interaction between the spins of carriers and localized holes in (8.3) and Hubbard U interaction for the localized holes in b_{1g}^* orbitals.

Then, in the case of a dopant hole with up-spin, the energy of b_{1g}^* state in a CuO_6 cluster with localized up-spin (A-site) is taken to be high so that the b_{1g}^* state at A-site is filled with holes even in undoped La_2CuO_4, while the energy of b_{1g}^* state in a CuO_6 cluster with localized down-spin (B-site) is low so that there are no holes in the b_{1g}^* state at B-site, i.e., the b_{1g}^* states at B-sites are empty. The difference between the energy of b_{1g}^* states at A-site and B-site is due to the strong U effect. Further, the energy of a_{1g}^* state at A-site is taken to be higher than that at B-site by Hund's coupling energy, while the energy of b_{1g} state at B-site is taken to be higher than that at A-site by the spin-singlet coupling in $^1A_{1g}$ state, so as to reproduce the characteristic electronic structure where up-spin carriers take the $^3B_{1g}$ state at A-site and the $^1A_{1g}$ state at B-site. In this chapter the energy of b_{1g}^*, b_{1g} or a_{1g}^* state indicates the energy for a electron but not a hole.

In this way one can include the many-body interaction effects of the Hubbard U interaction for the localized holes in b_{1g}^* orbital as well as of the exchange interaction in the K–S Hamiltonian (8.3) in the the 34 × 34 dimensional effective-interaction part $\tilde{H}_{\text{int}}(\boldsymbol{k})$. Further, all the matrix elements related to the transfer interactions which appear in the one-electron part of the 34 × 34 dimensional Hamiltonian matrix, $\tilde{H}_0(\boldsymbol{k})$, can be estimated from the Slater–Koster (SK) parameters. In the present calculation we use the values of the SK parameters fitted to an APW band calculation [152] by De Weert et al. [153] and thus the one-electron part of the Hamiltonian, $\tilde{H}_0(\boldsymbol{k})$, reproduces the APW bands well.

In order to obtain $\tilde{H}_{\text{int}}(\boldsymbol{k})$, we first construct the eigenstates localized at A-site or B-site by taking the linear combination of the doubly degenerated eigenstates of $\tilde{H}_0(\boldsymbol{k}_0)$, where vector $\boldsymbol{k}_0$ indicates $(\frac{\pi}{2a}, \frac{\pi}{2a}, 0)$ with a being the lattice constant in CuO$_2$ plane. The resultant eigenstates are $\sum_l \cos(\frac{\pi}{2a}x_l + \frac{\pi}{2a}y_l)\varphi_{al}$ and $\sum_l \sin(\frac{\pi}{2a}x_l + \frac{\pi}{2a}y_l)\varphi_{al}$, which we consider to be localized at A-site and B-site respectively, where φ_{al} are the Wannier type eigenstates of $\tilde{H}_0(\boldsymbol{k}_0)$.

If we take these functions as a basis function, the effective-interaction part $\tilde{H}_{\text{int}}(\boldsymbol{k})$ is obtained, by choosing the energy of the b$_{1g}^*$ state at A-site, that of b$_{1g}^*$ state at B-site, that of b$_{1g}$ state at B-site, that of a$_{1g}^*$ state at A-site and that of a$_{1g}^*$ state at B-site so as to reproduce the energy difference between multiplet ^{3}B$_{1g}$ and ^{1}A$_{1g}$ calculated by Kamimura and Eto [104]. Then by a unitary transformation we can obtain the expression of $\tilde{H}_{\text{int}}(\boldsymbol{k}_0)$ with the ordinary basis of Wannier type atomic functions.

The method described above is similar in its idea to the (LDA+U) method developed by Anisimov et al. [154] for copper oxides, but the interactions are treated accurately in the present method, while Anisimov et al. treated U as a disposal parameter. As described above, all the matrix elements in the 34×34 dimensional Hamiltonian matrix ($\tilde{H}$) become one-electron type, and thus we can diagonalize it easily. In this way we can obtain a band structure including the many-body effects in a molecular field approximation for LSCO.

10.2 Slater–Koster Method: Its Application to LSCO

The Slater–Koster method [155], in which the analytical form of the tight binding (TB) Hamiltonian is fitted to the first-principles band calculation, can be used to give insight into difficult problems which are intractable with a standard first-principles calculation method. Therefore, it has been used to consider the structural phase transition associated with a charge density wave [156], the phonon spectra and the electron–phonon mediated superconductivity in high T_c cuprates [157, 158]. In the present chapter we use the Slater–Koster (SK) method as a starting point for a many-body calculation of the electronic structure of LSCO. The SK method was first applied to LSCO by De Weert et al. [153]. They determined the on-site matrix elements and the overlap integrals so as to fit the analytical form of the tight binding Hamiltonian to the first-principles APW calculation. They performed the augmented-plane-wave (APW) calculation to generate the eigenvalues $E_n(\boldsymbol{k})$ and the angular momentum components, $Q_{nlm}(\boldsymbol{k})$, which mean the fraction of electronic charge in the nth band for the lth angular momentum component of the mth basis atom. In the Slater–Koster fits they identify the angular momentum components as the squares of the norms of the coefficients of TB wave functions in terms of atomic-like orbitals. In order to generate the TB band with a proper angular momentum character, they minimize the functional $F = \sum_{\boldsymbol{k},n}[f_n(\boldsymbol{k})]^2$, where

$$f_n(\boldsymbol{k}) = \left|E_n^{\mathrm{APW}}(\boldsymbol{k}) - E_n^{\mathrm{SK}}(\boldsymbol{k})\right| + \sum_{lm}\left|Q_{nlm}^{\mathrm{APW}}(\boldsymbol{k}) - Q_{nlm}^{\mathrm{SK}}(\boldsymbol{k})\right|/W \; ,$$

where the superscripts APW and SK denotes the first-principles calculated values and the Slater–Koster values, respectively, and W is a weight used to adjust the relative importance of $E_n(\boldsymbol{k})$ and $Q_{nlm}(\boldsymbol{k})$ in their fit. Thus the SK method affords a basis for a "tight binding" Hamiltonian as a starting point for many-body calculations. In this section we develop a formalism of a "tight binding" Hamiltonian for the undistorted crystal structure by using the Slater–Koster parameters.

In the tight-binding model, the Bloch functions are constructed from the atomic orbitals $\varphi_a(\boldsymbol{r} - \boldsymbol{R}_{l\mu})$ as

$$\Phi^0_{\mu a \boldsymbol{k}}(\boldsymbol{r}) = \frac{1}{\sqrt{N}} \sum_l e^{i\boldsymbol{k}\cdot\boldsymbol{R}_{l\mu}} \varphi_a(\boldsymbol{r} - \boldsymbol{R}_{l\mu}) \; , \tag{10.1}$$

where $\boldsymbol{R}_{l\mu} = \boldsymbol{R}_l + \boldsymbol{\tau}_\mu$ represents the position of the μth ion in the lth unit cell, $\boldsymbol{\tau}_\mu$ the position of the μth ion in the unit cell, N the total number of unit cells in the crystal, $\boldsymbol{k}$ a wave vector and a specifies an orbital.

Neglecting the overlap integrals, the energy eigenvalues and the wave functions are obtained by solving the following equation,

$$\mathrm{Det}|\tilde{H}_0(\boldsymbol{k}) - E^0_{n\boldsymbol{k}}\tilde{1}| = 0 \; , \tag{10.2}$$

where $\tilde{H}_0(\boldsymbol{k})$ is the Hamiltonian matrix and $\tilde{1}$ the unit matrix. The energy eigenvalues $E^0_{n\boldsymbol{k}}$ and the wave functions $\Psi^0_{n\boldsymbol{k}}(\boldsymbol{r})$ are represented by using the transformation matrix $\tilde{U}$

$$\tilde{E}_0(\boldsymbol{k}) = \tilde{U}^{-1}(\boldsymbol{k})\tilde{H}_0(\boldsymbol{k})\tilde{U}(\boldsymbol{k}) \tag{10.3}$$

$$\Psi^0_{n\boldsymbol{k}}(\boldsymbol{r}) = \sum_{\mu a} U_{\mu a, n}(\boldsymbol{k}) \Phi^0_{\mu a \boldsymbol{k}}(\boldsymbol{r}) \; , \tag{10.4}$$

where $\tilde{E}_0(\boldsymbol{k}) = E^0_{n\boldsymbol{k}}\tilde{1}$. The matrix elements of the Hamiltonian $\tilde{H}_0(\boldsymbol{k})$ is defined by

$$H^0_{\mu a \nu b}(\boldsymbol{k}) = \langle \Phi^0_{\mu a \boldsymbol{k}} | \mathcal{H}_e | \Phi^0_{\nu b \boldsymbol{k}} \rangle \; , \tag{10.5}$$

where $\mathcal{H}_e$ represents the one-electron Hamiltonian which may be regarded to include a part of electron correlation because the Slater–Koster (SK) parameters are determined so as to reproduce the electronic energy and the wave functions of first-principles band calculation. This Hamiltonian matrix $\tilde{H}_0(\boldsymbol{k})$ is expressed by taking the atomic orbitals as bases in the following way,

$$H^0_{\mu a \nu b}(\boldsymbol{k}) = \sum_{l-l'} e^{-i\boldsymbol{k}\cdot(\boldsymbol{R}_{l\mu}-\boldsymbol{R}_{l'\nu})} H^0_{l\mu a l'\nu b} \tag{10.6}$$

where

$$H^0_{l\mu a l'\nu b} = \langle \varphi_a(\boldsymbol{r} - \boldsymbol{R}_{l\mu})|\mathcal{H}_e|\varphi_b(\boldsymbol{r} - \boldsymbol{R}_{l'\nu})\rangle \ . \tag{10.7}$$

Further, all the matrix elements related to the transfer interactions which appear in the Hamiltonian matrix ($\tilde{H}_0(\boldsymbol{k})$) are expressed in terms of the SK parameters, which represent the transfer integrals between two atomic orbitals, c_m at the origin and c'_m at an arbitrary position $\boldsymbol{R}$, where c and c' represent s, p and d, and m denotes the magnetic quantum number of the orbital angular momentum with respect to the direction of $\boldsymbol{R}$. The SK parameters are conventionally symbolized as $t(cc'\sigma)$, $t(cc'\pi)$ and $t(cc'\delta)$, corresponding to $m = 0$, ± 1 and ± 2, respectively.

In the present treatment we restrict the basis functions to include $2p_x$, $2p_y$ and $2p_z$ atomic orbitals for each oxygen atom and $3d_{yz}$, $3d_{xz}$, $3d_{xy}$, $3d_{x^2-y^2}$ and $3d_{z^2}$ atomic orbitals for each Cu atom in the unit cell. Then the Hamiltonian matrix is expressed by 17 SK parameters if we consider only first neighbour interactions. They are listed in Table 10.1. In this table, for instance, $t(dd\sigma)$ represents the transfer integrals between two neighbouring Cu d orbitals with the magnetic quantum number $m = 0$ of the orbital angular momentum with respect to the Cu–Cu direction.

The Hamiltonian matrix is shown in Table 10.2, and the detailed expressions of its matrix elements are given in Sect. 10.5 "Appendix A" at the end of this chapter.

10.3 Computation Method to Calculate the Many-Electron Energy Bands: Its Application to LSCO

High-energy neutron scattering studies have shown a persistence of 2D antiferromagnetic spin correlation in the superconducting state of LSCO [159], and the ARPES results by Aebi et al. [160] have proved the prediction of a $\sqrt{2} \times \sqrt{2}$ antiferromagnetic local order. In this section we develop a computational method to calculate a new electronic structure in the superconducting concentration region based on the K–S model, in which, if the localized spins form AF ordering in a spin-correlated region, the carriers take the $^3B_{1g}$ high-spin multiplet state and the $^1A_{1g}$ low-spin multiplet state alternately in this spin-correlated region. In this respect a unit cell is taken so as to include two neighbouring CuO_6 octahedrons with localized up- and down-spins. This unit cell is called "antiferromagnetic (AF) unit cell", and two neighbouring CuO_6 octahedrons are called A-site and B-site, respectively.

The Hamiltonian matrix $\tilde{H}(\boldsymbol{k})$ consists of two parts; the one-electron part $\tilde{H}_0(\boldsymbol{k})$, and the effective-interaction part $\tilde{H}_{\text{int}}(\boldsymbol{k})$, as described in a previous section. In the AF unit cell, the one-electron part Hamiltonian matrix $\tilde{H}_0(\boldsymbol{k})$ is expressed by the following 34 × 34 matrix

$$\tilde{H}_0(\boldsymbol{k}) = \begin{bmatrix} \tilde{H}^0_{AA}(\boldsymbol{k}) & \tilde{H}^0_{BA}(\boldsymbol{k}) \\ \tilde{H}^0_{AB}(\boldsymbol{k}) & \tilde{H}^0_{BB}(\boldsymbol{k}) \end{bmatrix} \ , \tag{10.8}$$

Table 10.1. Slater–Koster parameters

[Figure: CuO$_6$ octahedron with Cu at origin, O(1)^1 and O(1)^2 in-plane oxygens along x and y, O(2)^1 and O(2)^2 apical oxygens along ±z]

On-site parameters

O(1): in plane	E_p^1
O(2): apical	E_p^2
Cu	E_{dxy}
	$E_{dx^2-y^2} = E_{dz^2}$

First-neighbor parameters

Cu–Cu	$t(dd\sigma)$
	$t(dd\pi)$
	$t(dd\delta)$
Cu–O(1)	$t_1(dp\sigma)$
	$t_1(dp\pi)$
Cu–O(2)	$t_2(dp\sigma)$
	$t_2(dp\pi)$
O(1)–O(1)	$t_1(pp\sigma)$
	$t_1(pp\pi)$
O(1)–O(2)	$t_2(pp\sigma)$
	$t_2(pp\pi)$
O(2)–O(2)	$t_3(pp\sigma)$
	$t_3(pp\pi)$

10.3 Computation Method to Calculate the Many-Electron Energy Bands

Table 10.2. Matrix elements of $\tilde{H}^0(\boldsymbol{k})$

		O(1)¹			O(1)²			O(2)¹			O(2)²			Cu				
		x	y	z	x	y	z	x	y	z	x	y	z	xy	yz	zx	x^2-y^2	z^2
O(1)¹	x	E_1	0	0	T_1	T_2	0	T_4	0	T_5	T_4^*	0	T_5^*	0	0	0	T_{13}	T_{14}
	y		E_1	0	T_2	T_1	0	0	T_6	0	0	T_6^*	0	T_{15}	0	0	0	0
	z			E_1	0	0	T_3	T_5	0	T_7	T_5^*	0	T_7^*	0	0	T_{15}	0	0
O(1)²	x				E_1	0	0	T_6'	0	0	$T_6'^*$	0	0	T_{15}'	0	0	0	0
	y					E_1	0	0	T_4'	T_5'	0	$T_4'^*$	$T_5'^*$	0	0	0	T_{13}'	T_{14}'
	z						E_1	0	T_5'	T_7'	0	$T_5'^*$	$T_7'^*$	0	T_{15}'	0	0	0
O(2)¹	x							E_2	0	0	T_8	T_{10}	T_{11}	0	0	$-T_{16}^*$	0	0
	y								E_2	0	T_{10}	T_8	T_{12}	0	$-T_{16}^*$	0	0	0
	z									E_2	T_{11}	T_{12}	T_9	0	0	0	0	$-T_{17}^*$
O(2)²	x										E_2	0	0	0	0	T_{16}	0	0
	y											E_2	0	0	T_{16}	0	0	0
	z												E_2	0	0	0	0	T_{17}
Cu	xy													E_3	0	0	0	0
	yz														E_4	0	0	0
	zx															E_5	0	0
	x^2-y^2																E_6	T_{18}
	z^2																	E_7

Table 10.3. The values of Slater–Koster parameters determined by De Weert et al.

		O(1)¹			O(1)²			O(2)¹			O(2)²			Cu				
		x	y	z	x	y	z	x	y	z	x	y	z	xy	yz	zx	x^2-y^2	z^2
O(1)¹	x	E_1	0	0	T_1	T_2	0	T_4	0	T_5	T_4^-	0	T_5^-	0	0	0	T_{13}	T_{14}
	y		E_1	0	T_2	T_1	0	0	T_6	0	0	T_6^-	0	T_{15}	0	0	0	0
	z			E_1	0	0	T_3	T_5	0	T_7	T_5^-	0	T_7^-	0	0	T_{15}	0	0
O(1)²	x				E_1	0	0	$T_6'^-$	0	0	$T_6'^-$	0	0	T_{15}'	0	0	0	0
	y					E_1	0	0	T_4'	T_5'	0	$T_4'^-$	$T_5'^-$	0	0	0	T_{13}'	T_{14}'
	z						E_1	0	T_5'	T_7'	0	$T_5'^-$	$T_7'^-$	0	T_{15}'	0	0	0
O(2)¹	x							E_2	0	0	T_8	T_{10}	T_{11}	0	0	$-T_{16}^-$	0	0
	y								E_2	0	T_{10}	T_8	T_{12}	0	$-T_{16}^-$	0	0	0
	z									E_2	T_{11}	T_{12}	T_9	0	0	0	0	$-T_{17}^-$
O(2)²	x										E_2	0	0	0	0	T_{16}	0	0
	y											E_2	0	0	T_{16}	0	0	0
	z												E_2	0	0	0	0	T_{17}
Cu	xy													E_3	0	0	0	0
	yz														E_4	0	0	0
	zx															E_5	0	0
	x^2-y^2																E_6	T_{18}
	z^2																	E_7

where $\tilde{H}^0_{AA}(\boldsymbol{k})$, $\tilde{H}^0_{BA}(\boldsymbol{k})$, $\tilde{H}^0_{AB}(\boldsymbol{k})$ and $\tilde{H}^0_{BB}(\boldsymbol{k})$ are the 17×17 matrices which represent Hamiltonian matrix elements between A- and A-sites, B- and A-sites, A- and B-sites, and B- and B-sites, respectively. These elements are defined in Table 10.3, and the expressions of these matrix elements are given in Sect. 10.5 "Appendix B" at the end of this chapter. In the present calculation we have used the values of the SK parameters fitted to the APW calculation [152] by De Weert et al. [153]. Those values are given in Table 10.4.

Now we take account of the many-body interaction terms of Hamiltonian (8.3) in the 34 × 34 dimensional effective-interaction part $\tilde{H}_{\mathrm{int}}(\boldsymbol{k})$. In order to include the effects of the exchange integrals between the spin of a dopant hole and localized spin, $K_{a_{1g}^*}$ and $K_{b_{1g}}$ in (8.3) and the Hubbard U-like parameter into the many-electron energy bands of a hole-carrier system, we first construct the antibonding b_{1g}^* orbital at A-site and B-site, mainly from a Cu $d_{x^2-y^2}$ atomic orbital, the bonding b_{1g} orbital at A-site and B-site from the O p_σ orbitals in a CuO$_2$ layer hybridized by a Cu $d_{x^2-y^2}$ atomic orbital, and

Table 10.4. Matrix elements of $\tilde{H}^0_{AA}(\boldsymbol{k})$, $\tilde{H}^0_{AB}(\boldsymbol{k})$, $\tilde{H}^0_{BA}(\boldsymbol{k})$ and $\tilde{H}^0_{BB}(\boldsymbol{k})$, where $T^- = T(-k_z)$

On-site parameters in Rydbergs		
O(1): in plane	E_p^1	0.2965
O(2): apical	E_p^2	0.3333
Cu	E_{dxy}	0.3506
	$E_{dx^2-y^2}$ E_{dz^2}	0.4375

First-neighbour parameters in Ry		
Cu–Cu	$t(dd\sigma)$	0.0048
	$t(dd\pi)$	−0.0049
	$t(dd\delta)$	−0.0058
Cu–O(1)	$t_1(dp\sigma)$	0.0921
	$t_1(dp\pi)$	0.0631
Cu–O(2)	$t_2(dp\sigma)$	0.0418
	$t_2(dp\pi)$	0.0277
O(1)–O(1)	$t_1(pp\sigma)$	0.0431
	$t_1(pp\pi)$	−0.0282
O(1)–O(2)	$t_2(pp\sigma)$	−0.0152
	$t_2(pp\pi)$	−0.0144
O(2)–O(2)	$t_3(pp\sigma)$	0.0126
	$t_3(pp\pi)$	−0.0018

a^*_{1g} orbital at A-site and B-site from Cu d_{z^2} orbital and O p_σ orbitals in a CuO_2 layer and O p_z orbitals of apical oxygen. The antibonding b^*_{1g} orbitals at A-site and B-site accommodate up-spin and down-spin holes, respectively, due to the Hubbard U interaction and the superexchange interaction. Then the a^*_{1g} orbital at A-site and the b_{1g} orbital at B-site participate in forming the $^3B_{1g}$ high-spin multiplet and the $^1A_{1g}$ low-spin multiplet, respectively, with the localized b^*_{1g} holes.

We construct localized states at A-site and B-site by taking a linear combination of the doubly degenerated eigenstates of the one-electron Hamiltonian $\tilde{H}_0(\boldsymbol{k}_0)$ where vector $\boldsymbol{k}_0$ indicates $(\frac{\pi}{2a}, \frac{\pi}{2a}, 0)$. This is possible because the eigenstates $|\boldsymbol{k}_0\rangle$ and $|-\boldsymbol{k}_0\rangle$ are degenerate, reflecting the fact that the difference between two wave vectors, $\boldsymbol{k}_0$ and $-\boldsymbol{k}_0$, coincides with a reciprocal lattice vector. The resultant eigenstates are $\sum_l \cos(\frac{\pi}{2a}x_l + \frac{\pi}{2a}y_l)\varphi_{al}$ and $\sum_l \sin(\frac{\pi}{2a}x_l + \frac{\pi}{2a}y_l)\varphi_{al}$, respectively, where φ_{al} are the Wannier type eigenstates of $\tilde{H}_0(\boldsymbol{k}_0)$, which are localized at the lth site and constructed with a linear combination of atomic orbitals. Strictly speaking, these eigenstates are not localized only at a particular site, but we consider these eigenstates as those localized at A-site and B-site. Using the transformation matrix $\tilde{U}(\boldsymbol{k}_0)$, which yields such localized eigenstates, $\tilde{H}_0(\boldsymbol{k}_0)$ is diagonalized;

$$\tilde{U}^{-1}(\boldsymbol{k}_0)\tilde{H}_0(\boldsymbol{k}_0)\tilde{U}(\boldsymbol{k}_0) = \tilde{E}_0(\boldsymbol{k}_0) . \tag{10.9}$$

10.3 Computation Method to Calculate the Many-Electron Energy Bands

The eigenstates of $\tilde{E}_0$ are expressed as linear combinations of atomic orbitals localized at A-site or B-site, such as a_{1g}^* orbital at A-site, a_{1g}^* orbital at B-site, b_{1g}^* orbital at A-site, b_{1g}^* orbital at B-site, and so on. In order to construct, for example, a_{1g}^* orbital at B-site, in numerical calculations we take a linear combination of two degenerate a_{1g}^* orbitals which correspond to the eigenstates $|\boldsymbol{k}_0\rangle$ and $|-\boldsymbol{k}_0\rangle$ so that the component of Cu d_{z^2} orbital at A-site disappears.

If we include the effects of the exchange interaction between the spin of a dopant hole and a localized spin, K in (8.3) into the carrier states, then, in the case of a dopant hole with up-spin, the energy of an electron occupying the a_{1g}^* state at A-site is taken to be higher than that at B-site by Hund's coupling energy, which is 2 eV [103]. On the other hand, as regards the energy of an electron occupying the b_{1g} state at B-site, it is first taken to be higher than that at A-site by the energy of the spin-singlet coupling in $^1A_{1g}$ multiplet which is 4 eV [103]. Then we have to proceed to include the effect of the crystalline potential in LSCO in the energy of b_{1g} state. The effect corresponds to the energy difference between the $^3B_{1g}$ and $^1A_{1g}$ multiplets due to the Madelung potential. According to the cluster calculation by Kamimura and Eto [103], this energy difference is found to be 2 eV. Thus 2 eV should be added to the on-site energy of the b_{1g} orbital, while leaving the on-site energy of the a_{1g}^* orbital remains unchanged. As a result the energy of b_{1g} state at B-site, which is the sum of the spin-singlet coupling energy, 4 eV, and the on-site energy of b_{1g} orbital, 2 eV, becomes 6 eV. Thus the up-spin carriers take the $^3B_{1g}$ state at A-site and the $^1A_{1g}$ state at B-site in the underdoped region. Lastly the energy of b_{1g}^* state in a CuO$_6$ cluster with localized up-spin (A-site) is taken to be higher than that in a CuO$_6$ cluster with localized down-spin (B-site) by the Hubbard U parameter, which is taken as 10 eV in the present treatment. Thus the localized spin band b_{1g}^* becomes separated from the hole carrier system.

Then the total Hamiltonian $\tilde{H}(\boldsymbol{k})$ is constructed with the one-electron part and the effective-interaction part, and the effective-interaction part has the eigenvalue of the b_{1g}^* state at A-site which is $+10$ eV, that of b_{1g}^* state at B-site -10 eV, that of b_{1g} state at B-site $+6$ eV, that of a_{1g}^* state at A-site $+1$ eV and that of a_{1g}^* state at B-site -1 eV. Here it should be noted that $\tilde{H}(\boldsymbol{k})$ is the Hamiltonian matrix for a electron but not a hole. Then the total Hamiltonian $\tilde{H}(\boldsymbol{k})$ should be transformed by transformation matrix $\tilde{U}(\boldsymbol{k}_0)$, as

$$\tilde{U}^{-1}(\boldsymbol{k}_0)\tilde{H}(\boldsymbol{k}_0)\tilde{U}(\boldsymbol{k}_0) = \tilde{E}_0(\boldsymbol{k}_0) + \tilde{E}_{\text{int}}(\boldsymbol{k}_0) , \qquad (10.10)$$

where

$$\tilde{E}_{\text{int}}(\boldsymbol{k}_0) = \begin{bmatrix} \ddots & & & & & & \\ & +10 & & & & & \\ & & -10 & & & & \\ & & & \ddots & & & \\ & & & & +1 & & \\ & & & & & -1 & \\ & & & & & & \ddots \\ & & & & & & & +6 \\ & & & & & & & & \ddots \end{bmatrix} \begin{array}{l} \vdots \\ \text{A-site } b_{1g}^* \\ \text{B-site } b_{1g}^* \\ \vdots \\ \text{A-site } a_{1g}^* \\ \text{B-site } a_{1g}^* \\ \vdots \\ \text{B-site } b_{1g} \\ \vdots \end{array} \quad (10.11)$$

with the energy being measured in eV. By inverse transformation we can obtain, $\tilde{H}_{\text{int}}(\boldsymbol{k}_0) = \tilde{U}(\boldsymbol{k}_0)\tilde{E}_{\text{int}}(\boldsymbol{k}_0)\tilde{U}^{-1}(\boldsymbol{k}_0)$. A similar calculation with respect to $\boldsymbol{k}_0' = (\frac{\pi}{2a}, -\frac{\pi}{2a}, 0)$ gives $\tilde{H}_{\text{int}}(\boldsymbol{k}_0')$ as well.

Then, using the approximation that the effective-interaction term $\tilde{H}_{\text{int}}$ has matrix elements only between nearest neighbour atomic orbitals, we can represent the $\boldsymbol{k}$ dependence of the effective-interaction part of the Hamiltonian matrix, $\tilde{H}_{\text{int}}(\boldsymbol{k})$, as is shown in Sect. 10.5 "Appendix C" at the end of this chapter. Then we can determine $\langle a|\tilde{H}_{\text{int}}^{\text{even}}|b\rangle$ and $\langle a|\tilde{H}_{\text{int}}^{\text{odd}}|b\rangle$, where "even" and "odd" mean that the interchanging of the two atomic orbitals, a and b, produces $+1$ and -1 in sign, respectively. In this way we can include the exchange interaction terms of the K–S Hamiltonian (8.3) and the Hubbard U interaction for the localized holes in b_{1g}^* orbital in the the 34×34 dimensional effective-interaction part $\tilde{H}_{\text{int}}(\boldsymbol{k})$. As for the value of the difference between $\epsilon_{a_{1g}^*}$ and $\epsilon_{b_{1g}}$ in (8.3), it is taken so as to reproduce the energy difference between multiplets $^3B_{1g}$ and $^1A_{1g}$ calculated by Kamimura and Eto [104].

As described above, all the matrix elements in the 34×34 dimensional Hamiltonian matrix ($\tilde{H}$) become of one-electron type as the result of the mean field approximation, and thus we can diagonalize it easily. In this way we can obtain a band structure including the many-body effects, which is treated as a molecular field acting on the dopant holes for LSCO. In the following we will present the results of the many-body included energy band for LSCO calculated by the computational method described in the present chapter.

10.4 Computation Method Applied to YBCO Materials

So far we have described a method for deriving the energy bands including the many-body effects based on the K–S model following Ushio and Kamimura, and applied it to LSCO. In this section we will apply the present method to YBCO$_7$, following calculations by Nomura and Kamimura [105]. In the

10.4 Computation Method Applied to YBCO Materials

case of LSCO, Ushio and Kamimura expressed the first-principles augmented-plane-wave (APW) or linearized-augmented-plane-wave (LAPW) band structure of La_2CuO_4 in terms of a tight-binding (TB) band structure following De Weert et al. [153], and calculated various physical properties such as electron–phonon interaction, Hall coefficient, resistivity, etc.

Thus such TB parametrization is an important tool for calculating a number of physical properties. In this context Nomura and Kamimura performed the TB parametrization for $YBCO_7$. Although the TB parametrization was already done by De Weert et al. for $YBCO_7$, Nomura and Kamimura found that wavefunctions corresponding to each TB energy band by De Weert et al. are not consistent with those obtained by APW or LAPW band structure calculated for $YBCO_7$. In this context Nomura and Kamimura performed newly the TB parametrization for the energy bands numerically calculated for $YBCO_7$ to reproduce not only the energy band shape but also wavefunctions for each band. In this section we describe their method with regard to the Slater–Koster fits for $YBa_2Cu_3O_7$.

As we described in previous sections, the Slater–Koster (SK) method [155], which treats TB matrix elements and overlap integrals as disposable parameters to be determined by fitting the TB band structure to first-principles calculated energy bands, can be used to give insight into difficult problems which are intractable with a standard first-principles calculation method. In the case of $YBCO_7$, Krakauer et al. [161] performed LAPW calculations to generate eigenvalues $E_n(k)$ and angular momentum components $Q_{nlm}(k)$ of an energy band. Here $Q_{nlm}(k)$ means the fraction of electronic charge in the nth band for the lth angular momentum component of the mth basis atom. This quantity is used to decompose the density of states. Then De Weert et al. determined the SK parameters to reproduce the bands presented by Krakauer et al. near the Fermi level.

They omitted Ba atoms and restricted the basis to Y-d, Cu-d, and O-p states, obtaining a 41×41 secular equation. They considered first-, second-, and third-neighbour hopping elements, so that the TB fit required 79 SK parameters. The CuO_2 planes consist of sites denoted Cu(2), O(2), and O(3), and the chain atoms are denoted as Cu(1) and O(1). The O(4) sites lie between chain and plane copper atoms, but are much closer to the chain Cu(1) sites. In this compound, none of the atoms sit at sites of local cubic or even tetragonal symmetry. Thus, all the p and d bands have, in principle, crystal-field splittings. This is particularly important for Cu(1), which has a very asymmetric local environment. Consequently, De Weert et al. described the O-p on-site energies with three distinct values, and the Cu-d on-site energies with five distinct values. The coordinates of the atoms they used are given in Table 10.5, and the neighbour distances they used are in Table 10.6. The structure is orthorhombic, with 13 atoms per unit cell distributed among eight distinct sites. The following lattice constants are adopted for $YBCO_7$ [153], $a = 7.2249$ a.u., $b = 1.01655$ a.u., and $c = 3.05599$ a.u.

Table 10.5. Structure of $YBa_2Cu_3O_7$. $a = 7.2249$ a.u., $b = 1.01655a$, $c = 3.05599a$

Atom	x(unit of a)	y(unit of b)	z(unit of c)	
Y	0.500	0.500	0.500	
Ba	0.500	0.500	0.1846	
Ba	0.500	0.500	0.8154	
Cu(1)	0.000	0.000	0.000	Chain
Cu(2)	0.000	0.000	0.3551	Plane
Cu(2)	0.000	0.000	0.6449	Plane
O(1)	0.000	0.500	0.000	Chain
O(2)	0.500	0.000	0.3781	Plane
O(2)	0.500	0.000	0.6219	Plane
O(3)	0.000	0.500	0.3779	Plane
O(3)	0.000	0.500	0.6221	Plane
O(4)	0.000	0.000	0.1579	
O(4)	0.000	0.000	0.8421	

Table 10.6. Yttrium, copper, and oxygen neighbours for $YBa_2Cu_3O_7$ (units of $a = 7.2249$ a.u.). ($*$ indicates that parameters for neighbours at $b = 1.01655a$ were the same as for this neighbour.)

First neighbours

	Cu(1)	Cu(2)	O(1)	O(2)	O(3)	O(4)
Y	$\cdots$	0.8393	$\cdots$	0.6301	0.6305	$\cdots$
Cu(1)	1.000*	$\cdots$	0.5083	$\cdots$	$\cdots$	0.4825
Cu(2)		0.8856	$\cdots$	0.5049	0.5048	0.6026
O(1)			1.000*	$\cdots$	$\cdots$	0.7006
O(2)				0.7463	0.7130	0.8384
O(3)					0.7463	0.8427
O(4)						0.9651

Second neighbours

	Cu(2)	O(2)	O(3)	O(4)
Cu(2)	1.000*	0.9564	0.9613	$\cdots$
O(2)		1.000*	$\cdots$	$\cdots$
O(3)			1.000*	$\cdots$
O(4)				1.000*

Third neighbours

	O(2)	O(3)
Cu(2)	1.1350	1.1239

10.4 Computation Method Applied to YBCO Materials

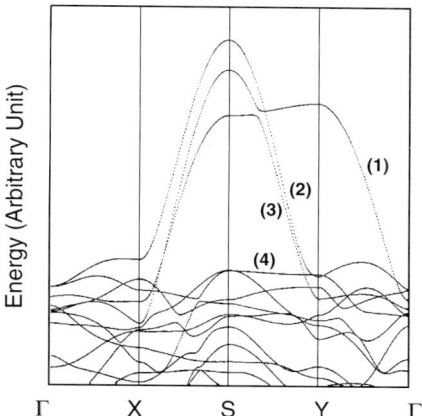

Fig. 10.1. The energy bands in $YBa_2Cu_3O_7$ near the Fermi level

The shape of the bands presented by De Weert et al. [153] shows excellent agreement near the Fermi level compared with that presented by Krakauer et al. by LAPW calculations. However, by calculating the character of wavefunctions for several energy bands at and just below the Fermi level, Nomura and Kamimura found that the characters of wavefunctions in the energy bands by De Weert et al. are completely different from the results calculated by Yu et al. [148], Andersen et al. [162], and so on. Figure 10.1 shows the energy bands for $YBCO_7$ near the Fermi level. According to Yu et al. or Andersen et al., the number 1 band, enclosed with a circle in Fig. 10.1, consists mainly of $Cu(1)d_{y^2-z^2}$-$O(1)p_y$-$O(4)p_z$ orbitals, the number 2 and 3 bands consist mainly of $Cu(2)d_{x^2-y^2}$-$O(2)p_x$-$O(3)p_y$ orbitals, and the number 4 band consists mainly of $Cu(1)d_{yz}$-$O(1)p_z$-$O(4)p_y$ orbitals, while according to the TB bands presented by De Weert et al., the number 1 band in Fig. 10.1 consists mainly of $Cu(1)d_{yz}$-$O(1)p_z$-$O(4)p_y$ orbitals, the number 2 and 3 bands consist mainly of $Cu(2)d_{xy}$-$O(2)p_y$-$O(3)p_x$ orbitals, and the number 4 band consists mainly of $Cu(1)d_{xy}$-$O(1)p_x$ orbitals. Thus, we have to say that the SK parameters in $YBCO_7$ determined by De Weert et al. are not appropriate. In this context, Nomura and Kamimura redetermined the SK parameters for $YBCO_7$ in such a way as to fit the character of wavefunctions to that estimated by Yu et al. [148], Andersen et al. [162], and so on, as well as the shape of the present TB bands to that presented by Krakauer et al. near the Fermi level.

In the calculation of the TB bands by SK method, Nomura and Kammimura omitted Y and Ba atoms, and restricted the bases only to Cu-d and O-p states. As a result, a 36×36 secular equation for the TB Hamiltonian is obtained. In their calculation they considered first-, second-, and third-neighbour hopping elements, so that the fit of the TB bands to the bands numerically calculated by Krakauer et al. required 71 SK parameters.

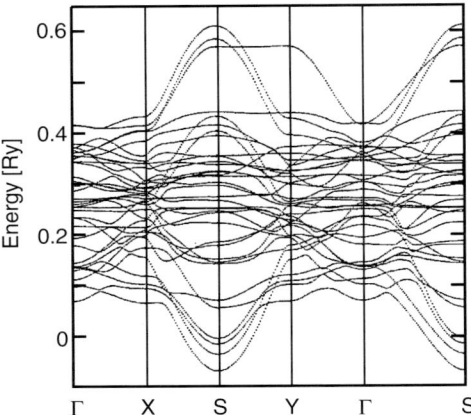

Fig. 10.2. The bands for YBa$_2$Cu$_3$O$_7$ by using the SK parameters determined by us

The coordinates of the atoms they used are given in Table 10.5, and the neighbour distances they used are given in Table 10.6. The SK parameters determined by them are given in Table 10.7, and the fitted bands are shown in Fig. 10.2. The Fermi level lies at 0.442 Ry, which is almost exactly equal to the LAPW value, 0.442 Ry [161]. The number 1 band, corresponding to the number enclosed with a circle in Fig. 10.1, consists mainly of Cu(1)$d_{x^2-y^2}$-Cu(1)d_{z^2}-O(1)p_y-O(4)p_z orbitals, the number 2 and 3 bands consist mainly of Cu(2)$d_{x^2-y^2}$-O(2)p_x-O(3)p_y orbitals, and the number 4 band consists mainly of Cu(1)d_{yz}-O(1)p_z-O(4)p_y orbitals. These are consistent with the results obtained by Yu et al., Andersen et al., and others.

Let us explain in more detail about what is the most serious difference between the SK parameters determined by Nomura and Kamimura and those presented by De Weert et al. In the SK parameters of De Weert et al., the on-site parameters for Cu(1)d_{yz}, O(2)p_y and O(3)p_x which are 0.3956, 0.2909, 0.3420 Ry, respectively, and the first neighbour parameters for Cu(1)-O(1)$pd\pi$, Cu(2)-O(2)$pd\pi$ and Cu(2)-O(3)$pd\pi$, which are 0.1035, 0.0842 and 0.0565 Ry in absolute values, respectively, are much larger than those listed in Table 10.7. In particular, as to the SK parameters which lead to the energy bands at and above the Fermi level, those related to π character are larger than those related to σ character in De Weert et al.'s paper, just opposite to the trend in this chapter. This is a reason why the wavefunctions of energy bands numbered 1, 2, 3 and 4 calculated by the SK parameters of De Weert et al. are different from those of Yu et al., of Anderson et al. and also of Nomura and Kamimura.

By the SK method, Nomura and Kamimumra determined 71 SK parameters by fitting the TB bands to both numerically-calculated energy bands and wavefunctions. In their calculation they also showed that the SK parameters

10.4 Computation Method Applied to YBCO Materials

Table 10.7. The SK parameters for $YBa_2Cu_3O_7$

On-site parameters

Cu(1):	d_{xy}	0.2169	O(1):	p_x	0.2530
	d_{yz}	0.3082		p_y	0.3071
	d_{zx}	0.3349		p_z	0.2878
	$d_{x^2-y^2}$	0.3525	O(2):	p_x	0.2700
	d_{z^2}	0.3516		p_y	0.2000
Cu(2):	d_{xy}	0.2500		p_z	0.2700
	d_{yz}	0.3124	O(3):	p_x	0.2000
	d_{zx}	0.2737		p_y	0.2700
	$d_{x^2-y^2}$	0.2500		p_z	0.2700
	d_{z^2}	0.3100	O(4):	p_x	0.1931
				p_y	0.3268
				p_z	0.2494

First-neighbour parameters

Cu(1)–Cu(1):	$dd\sigma$	-0.0200	O(1)–O(1):	$pp\sigma$	0.0254
	$dd\pi$	0.0008		$pp\pi$	0.0119
	$dd\delta$	-0.0020	O(1)–O(4):	$pp\sigma$	0.0123
Cu(2)–Cu(2):	$dd\sigma$	-0.0060		$pp\pi$	0.0341
	$dd\pi$	0.0035	O(2)–O(2):	$pp\sigma$	0.0350
	$dd\delta$	-0.0030		$pp\pi$	-0.0145
Cu(1)–O(1):	$pd\sigma$	-0.1173	O(2)–O(3):	$pp\sigma$	0.0830
	$pd\pi$	0.0615		$pp\pi$	-0.0330
Cu(1)–O(4):	$pd\sigma$	-0.0500	O(2)–O(4):	$pp\sigma$	0.0200
	$pd\pi$	0.0353		$pp\pi$	-0.0100
Cu(2)–O(2):	$pd\sigma$	-0.0800	O(3)–O(3):	$pp\sigma$	0.0250
	$pd\pi$	0.0150		$pp\pi$	-0.0145
Cu(2)–O(3):	$pd\sigma$	-0.0750	O(3)–O(4):	$pp\sigma$	0.0300
	$pd\pi$	0.0150		$pp\pi$	-0.0150
Cu(2)–O(4):	$pd\sigma$	-0.0300	O(4)–O(4):	$pp\sigma$	0.0300
	$pd\pi$	0.0140		$pp\pi$	-0.0050

Second-neighbour parameters

Cu(2)–Cu(2):	$dd\sigma$	-0.0020	O(2)–O(2):	$pp\sigma$	0.0100
	$dd\pi$	0.0012		$pp\pi$	-0.0050
	$dd\delta$	-0.0002	O(3)–O(3):	$pp\sigma$	0.0100
Cu(2)–O(2):	$pd\sigma$	-0.0100		$pp\pi$	-0.0050
	$pd\pi$	0.0045	O(4)–O(4):	$pp\sigma$	0.0080
Cu(2)–O(3):	$pd\sigma$	-0.0100		$pp\pi$	-0.0150
	$pd\pi$	0.0045			

Third-neighbour parameters

Cu(2)–O(2):	$pd\sigma$	-0.0025	Cu(2)–O(3):	$pd\sigma$	-0.0025
	$pd\pi$	0.0010		$pd\pi$	0.0010

previously determined by De Weert et al. do not reproduce the wavefunctions of the bands near the Fermi level so that their values for the SK parameters are not appropriate. In this respect we would like to make a further remark as to the SK parameters by De Weert et al..

Following Kamimura and Ushio and using the results of the cluster calculations by the Kamimura and Sano, Nomura and Kamimumra calculated the many-electron energy bands, the Fermi surfaces and density of states on the basis of the K–S Hamiltonian for the Kamimura–Suwa model. Here we would like to point out that the Fermi surfaces calculated by the SK parameters of De Weert et al. do not have closed surfaces for carriers originated from CuO_2 planes and are extremely different from those observed by experiments, while the Fermi surfaces obtained by other groups are closed shapes although there is difference in their size, small and large.

10.5 Appendix

In this section the detailed expressions of the matrix elements in Tables 10.2, 10.3 and $\tilde{H}_{\text{int}}$ in Sect. 10.3 are given in Appendices A, B and C, respectively.

Appendix A

The matrix elements of the Hamiltonian matrix shown in Table 10.2, are expressed with 17 SK parameters as

$$E_1 = E_p^1$$
$$E_2 = E_p^2$$
$$E_3 = E_{dxy} + 2\, t(dd\pi)(\cos k_x a + \cos k_y a)$$
$$E_4 = E_{dxy} + 2\, t(dd\pi)\cos k_y a + 2\, t(dd\delta)\cos k_x a$$
$$E_5 = E_{dxy} + 2\, t(dd\pi)\cos k_x a + 2\, t(dd\delta)\cos k_y a$$
$$E_6 = E_{dx^2-y^2} + \left[\frac{3}{2} t(dd\sigma) + \frac{1}{2} t(dd\delta)\right](\cos k_x a + \cos k_y a)$$
$$E_7 = E_{dx^2-y^2} + \left[\frac{1}{2} t(dd\sigma) + \frac{3}{2} t(dd\delta)\right](\cos k_x a + \cos k_y a)$$

$$T_1 = 2\,[\,t_1(pp\sigma) + t_1(pp\pi)\,]\,\cos\frac{k_x a}{2}\cos\frac{k_y a}{2}$$
$$T_2 = -2\,[\,t_1(pp\sigma) - t_1(pp\pi)\,]\,\sin\frac{k_x a}{2}\sin\frac{k_y a}{2}$$
$$T_3 = 4\,t_1(pp\pi)\cos\frac{k_x a}{2}\cos\frac{k_y a}{2}$$

$$T_4 = 2\,[\,l_1^2\,t_2(pp\sigma) + (1-l_1^2)\,t_2(pp\pi)\,]\,\cos\frac{k_x a}{2}e^{i\,0.364 k_z c/2}$$

$$T_5 = i\,2l_1 n_1\,[\,t_2(pp\sigma) - t_2(pp\pi)\,]\,\sin\frac{k_x a}{2}e^{i\,0.364 k_z c/2}$$

$$T_6 = 2\,t_2(pp\pi)\cos\frac{k_x a}{2}e^{i\,0.364 k_z c/2}$$

$$T_7 = 2\,[\,n_1^2\,t_2(pp\sigma) + (1-n_1^2)\,t_2(pp\pi)\,]\,\cos\frac{k_x a}{2}e^{i\,0.364 k_z c/2}$$

$$T_4' = 2\,[\,l_1^2\,t_2(pp\sigma) + (1-l_1^2)\,t_2(pp\pi)\,]\,\cos\frac{k_y a}{2}e^{i\,0.364 k_z c/2}$$

$$T_5' = i\,2l_1 n_1\,[\,t_2(pp\sigma) - t_2(pp\pi)\,]\,\sin\frac{k_y a}{2}e^{i\,0.364 k_z c/2}$$

$$T_6' = 2\,t_2(pp\pi)\cos\frac{k_y a}{2}e^{i\,0.364 k_z c/2}$$

$$T_7' = 2\,[\,n_1^2\,t_2(pp\sigma) + (1-n_1^2)\,t_2(pp\pi)\,]\,\cos\frac{k_y a}{2}e^{i\,0.364 k_z c/2}$$

$$T_8 = 4\,[\,l_2^2\,t_3(pp\sigma) + (1-l_2^2)\,t_3(pp\pi)\,]\,\cos\frac{k_x a}{2}\cos\frac{k_y a}{2}e^{i\,0.272 k_z c/2}$$

$$T_9 = 4\,[\,n_2^2\,t_3(pp\sigma) + (1-n_2^2)\,t_3(pp\pi)\,]\,\cos\frac{k_x a}{2}\cos\frac{k_y a}{2}e^{i\,0.272 k_z c/2}$$

$$T_{10} = -4l_2^2\,[\,t_3(pp\sigma) - t_3(pp\pi)\,]\,\sin\frac{k_x a}{2}\sin\frac{k_y a}{2}e^{i\,0.272 k_z c/2}$$

$$T_{11} = i\,4l_2 n_2\,[\,t_3(pp\sigma) - t_3(pp\pi)\,]\,\sin\frac{k_x a}{2}\cos\frac{k_y a}{2}e^{i\,0.272 k_z c/2}$$

$$T_{12} = i\,4l_2 n_2\,[\,t_3(pp\sigma) - t_3(pp\pi)\,]\,\cos\frac{k_x a}{2}\sin\frac{k_y a}{2}e^{i\,0.272 k_z c/2}$$

$$T_{13} = i\,\sqrt{3}\,t_1(pd\sigma)\sin\frac{k_x a}{2}$$

$$T_{14} = -i\,t_1(pd\sigma)\sin\frac{k_x a}{2}$$

$$T_{15} = i\,2\,t_1(pd\pi)\sin\frac{k_x a}{2}$$

$$T_{13}' = -i\,\sqrt{3}\,t_1(pd\sigma)\sin\frac{k_y a}{2}$$

$$T_{14}' = -i\,t_1(pd\sigma)\sin\frac{k_y a}{2}$$

$$T_{15}' = i\,2\,t_1(pd\pi)\sin\frac{k_y a}{2}$$

$$T_{16} = t_2(pd\pi)e^{i\,0.364 k_z c/2}$$

$$T_{17} = t_2(pd\sigma)e^{i\,0.364 k_z c/2}$$

$$T_{18} = \frac{\sqrt{3}}{2}\,[\,-t(dd\sigma) + t(dd\delta)\,]\,(\cos k_x a - \cos k_y a) \tag{10.12}$$

where

$$l_1 = \frac{0.5a}{\sqrt{(0.5a)^2 + (0.364c/2)^2}}$$

$$n_1 = \frac{0.364c/2}{\sqrt{(0.5a)^2 + (0.364c/2)^2}}$$

$$l_2 = \frac{0.5a}{\sqrt{2(0.5a)^2 + (0.272c/2)^2}}$$

$$n_2 = \frac{0.272c/2}{\sqrt{2(0.5a)^2 + (0.272c/2)^2}} \tag{10.13}$$

Appendix B

The matrix elements in Table 10.3 are expressed with SK parameters as follows. The on-site elements of $\tilde{H}^0_{AA}(\boldsymbol{k})$ and $\tilde{H}^0_{BB}(\boldsymbol{k})$ are given below;

$$\begin{aligned}
E_1 &= E_p^1 \\
E_2 &= E_p^2 \\
E_3 &= E_{dxy} \\
E_4 &= E_{dxy} \\
E_5 &= E_{dxy} \\
E_6 &= E_{dx^2-y^2} \\
E_7 &= E_{dx^2-y^2} ,
\end{aligned} \tag{10.14}$$

The diagonal elements of $\tilde{H}^0_{AB}(\boldsymbol{k})$ and $\tilde{H}^0_{BA}(\boldsymbol{k})$ are given below;

$$\begin{aligned}
E_1 &= 0 \\
E_2 &= 0 \\
E_3 &= +2\ t(dd\pi)(\cos k_x a + \cos k_y a) \\
E_4 &= +2\ t(dd\pi)\cos k_y a + 2\ t(dd\delta)\cos k_x a \\
E_5 &= +2\ t(dd\pi)\cos k_x a + 2\ t(dd\delta)\cos k_y a \\
E_6 &= + \left[\frac{3}{2} t(dd\sigma) + \frac{1}{2} t(dd\delta)\right] (\cos k_x a + \cos k_y a) \\
E_7 &= + \left[\frac{1}{2} t(dd\sigma) + \frac{3}{2} t(dd\delta)\right] (\cos k_x a + \cos k_y a) ,
\end{aligned} \tag{10.15}$$

Finally the off-diagonal elements of $\tilde{H}^0_{AA}(\boldsymbol{k})$ and $\tilde{H}^0_{BB}(\boldsymbol{k})$ are given below;

$$T_1 = \frac{1}{2} [\ t_1(pp\sigma) + t_1(pp\pi)\ ]\ (e^{i\frac{k_x a}{2}} e^{-i\frac{k_y a}{2}} + e^{-i\frac{k_x a}{2}} e^{i\frac{k_y a}{2}})$$

$$T_2 = -\frac{1}{2}[\,t_1(pp\sigma) - t_1(pp\pi)\,]\,(e^{i\frac{k_x a}{2}}e^{-i\frac{k_y a}{2}} + e^{-i\frac{k_x a}{2}}e^{i\frac{k_y a}{2}})$$

$$T_3 = t_1(pp\pi)(e^{i\frac{k_x a}{2}}e^{-i\frac{k_y a}{2}} + e^{-i\frac{k_x a}{2}}e^{i\frac{k_y a}{2}})$$

$$T_4 = [\,l_1^2\,t_2(pp\sigma) + (1-l_1^2)\,t_2(pp\pi)\,]\,e^{-i\frac{k_x a}{2}}e^{i\,0.364 k_z c/2}$$

$$T_5 = -l_1 n_1\,[\,t_2(pp\sigma) - t_2(pp\pi)\,]\,e^{-i\frac{k_x a}{2}}e^{i\,0.364 k_z c/2}$$

$$T_6 = t_2(pp\pi)e^{-i\frac{k_x a}{2}}e^{i\,0.364 k_z c/2}$$

$$T_7 = [\,n_1^2\,t_2(pp\sigma) + (1-n_1^2)\,t_2(pp\pi)\,]\,e^{-i\frac{k_x a}{2}}e^{i\,0.364 k_z c/2}$$

$$T_4' = [\,l_1^2\,t_2(pp\sigma) + (1-l_1^2)\,t_2(pp\pi)\,]\,e^{-i\frac{k_y a}{2}}e^{i\,0.364 k_z c/2}$$

$$T_5' = -l_1 n_1\,[\,t_2(pp\sigma) - t_2(pp\pi)\,]\,e^{-i\frac{k_y a}{2}}e^{i\,0.364 k_z c/2}$$

$$T_6' = t_2(pp\pi)e^{-i\frac{k_y a}{2}}e^{i\,0.364 k_z c/2}$$

$$T_7' = [\,n_1^2\,t_2(pp\sigma) + (1-n_1^2)\,t_2(pp\pi)\,]\,e^{-i\frac{k_y a}{2}}e^{i\,0.364 k_z c/2}$$

$$T_8 = [\,l_2^2\,t_3(pp\sigma) + (1-l_2^2)\,t_3(pp\pi)\,](e^{i\frac{k_x a}{2}}e^{-i\frac{k_y a}{2}} + e^{-i\frac{k_x a}{2}}e^{i\frac{k_y a}{2}})e^{i\,0.272 k_z c/2}$$

$$T_9 = [\,n_2^2 t_3(pp\sigma) + (1-n_2^2)\,t_3(pp\pi)\,]\,(e^{i\frac{k_x a}{2}}e^{-i\frac{k_y a}{2}} + e^{-i\frac{k_x a}{2}}e^{i\frac{k_y a}{2}})e^{i\,0.272 k_z c/2}$$

$$T_{10} = -l_2^2\,[\,t_3(pp\sigma) - t_3(pp\pi)\,]\,(e^{i\frac{k_x a}{2}}e^{-i\frac{k_y a}{2}} + e^{-i\frac{k_x a}{2}}e^{i\frac{k_y a}{2}})e^{i\,0.272 k_z c/2}$$

$$T_{11} = l_2 n_2\,[\,t_3(pp\sigma) - t_3(pp\pi)\,]\,(e^{i\frac{k_x a}{2}}e^{-i\frac{k_y a}{2}} - e^{-i\frac{k_x a}{2}}e^{i\frac{k_y a}{2}})e^{i\,0.272 k_z c/2}$$

$$T_{12} = l_2 n_2\,[t_3(pp\sigma) - t_3(pp\pi)\,]\,(-e^{i\frac{k_x a}{2}}e^{-i\frac{k_y a}{2}} + e^{-i\frac{k_x a}{2}}e^{i\frac{k_y a}{2}})e^{i\,0.272 k_z c/2}$$

$$T_{13} = -\frac{\sqrt{3}}{2}\,t_1(pd\sigma)e^{-i\frac{k_x a}{2}}$$

$$T_{14} = \frac{1}{2}t_1(pd\sigma)e^{-i\frac{k_x a}{2}}$$

$$T_{15} = -\,t_1(pd\pi)e^{-i\frac{k_x a}{2}}$$

$$T_{13}' = \frac{\sqrt{3}}{2}\,t_1(pd\sigma)e^{-i\frac{k_y a}{2}}$$

$$T_{14}' = \frac{1}{2}t_1(pd\sigma)e^{-i\frac{k_y a}{2}}$$

$$T_{15}' = -\,t_1(pd\pi)e^{-i\frac{k_y a}{2}}$$

$$T_{16} = t_2(pd\pi)e^{i\,0.364 k_z c/2}$$

$$T_{17} = t_2(pd\sigma)e^{i\,0.364 k_z c/2}$$

$$T_{18} = 0\,, \tag{10.16}$$

and for the off-diagonal elements of $\tilde{H}^0_{AB}(\boldsymbol{k})$ and $\tilde{H}^0_{BA}(\boldsymbol{k})$

$$T_1 = \frac{1}{2}\,[\,t_1(pp\sigma) + t_1(pp\pi)\,]\,(e^{i\frac{k_x a}{2}}e^{i\frac{k_y a}{2}} + e^{-i\frac{k_x a}{2}}e^{-i\frac{k_y a}{2}})$$

$$T_2 = \frac{1}{2}\,[\,t_1(pp\sigma) - t_1(pp\pi)\,]\,(e^{i\frac{k_x a}{2}}e^{i\frac{k_y a}{2}} + e^{-i\frac{k_x a}{2}}e^{-i\frac{k_y a}{2}})$$

$$T_3 = t_1(pp\pi)\cos\frac{k_x a}{2}(e^{i\frac{k_x a}{2}}e^{i\frac{k_y a}{2}} + e^{-i\frac{k_x a}{2}}e^{-i\frac{k_y a}{2}})$$

$$T_4 = [\, l_1^2\, t_2(pp\sigma) + (1-l_1^2)\, t_2(pp\pi)\,]\, e^{i\frac{k_x a}{2}} e^{i\, 0.364 k_z c/2}$$

$$T_5 = l_1 n_1\, [\, t_2(pp\sigma) - t_2(pp\pi)\,]\, e^{i\frac{k_x a}{2}} e^{i\, 0.364 k_z c/2}$$

$$T_6 = t_2(pp\pi) e^{i\frac{k_x a}{2}} e^{i\, 0.364 k_z c/2}$$

$$T_7 = [\, n_1^2\, t_2(pp\sigma) + (1-n_1^2)\, t_2(pp\pi)\,]\, e^{i\frac{k_x a}{2}} e^{i\, 0.364 k_z c/2}$$

$$T_4' = [\, l_1^2\, t_2(pp\sigma) + (1-l_1^2)\, t_2(pp\pi)\,]\, e^{i\frac{k_y a}{2}} e^{i\, 0.364 k_z c/2}$$

$$T_5' = l_1 n_1\, [\, t_2(pp\sigma) - t_2(pp\pi)\,]\, e^{i\frac{k_y a}{2}} e^{i\, 0.364 k_z c/2}$$

$$T_6' = t_2(pp\pi) e^{i\frac{k_y a}{2}} e^{i\, 0.364 k_z c/2}$$

$$T_7' = [\, n_1^2\, t_2(pp\sigma) + (1-n_1^2)\, t_2(pp\pi)\,]\, e^{i\frac{k_y a}{2}} e^{i\, 0.364 k_z c/2}$$

$$T_8 = [\, l_2^2\, t_3(pp\sigma) + (1-l_2^2)\, t_3(pp\pi)\,]\, (e^{i\frac{k_x a}{2}} e^{i\frac{k_y a}{2}} + e^{-i\frac{k_x a}{2}} e^{-i\frac{k_y a}{2}}) e^{i\, 0.364 k_z c/2}$$

$$T_9 = [\, n_2^2\, t_3(pp\sigma) + (1-n_2^2)\, t_3(pp\pi)\,]\, (e^{i\frac{k_x a}{2}} e^{i\frac{k_y a}{2}} + e^{-i\frac{k_x a}{2}} e^{-i\frac{k_y a}{2}}) e^{i\, 0.272 k_z c/2}$$

$$T_{10} = l_2^2\, [\, t_3(pp\sigma) - t_3(pp\pi)\,]\, (e^{i\frac{k_x a}{2}} e^{i\frac{k_y a}{2}} + e^{-i\frac{k_x a}{2}} e^{-i\frac{k_y a}{2}}) e^{i\, 0.272 k_z c/2}$$

$$T_{11} = l_2 n_2\, [\, t_3(pp\sigma) - t_3(pp\pi)\,]\, (e^{i\frac{k_x a}{2}} e^{i\frac{k_y a}{2}} - e^{-i\frac{k_x a}{2}} e^{-i\frac{k_y a}{2}}) e^{i\, 0.272 k_z c/2}$$

$$T_{12} = l_2 n_2\, [\, t_3(pp\sigma) - t_3(pp\pi)\,]\, (e^{i\frac{k_x a}{2}} e^{i\frac{k_y a}{2}} - e^{-i\frac{k_x a}{2}} e^{-i\frac{k_y a}{2}}) e^{i\, 0.272 k_z c/2}$$

$$T_{13} = \frac{\sqrt{3}}{2}\, t_1(pd\sigma) e^{i\frac{k_x a}{2}}$$

$$T_{14} = -\frac{1}{2} t_1(pd\sigma) e^{i\frac{k_x a}{2}}$$

$$T_{15} = t_1(pd\pi) e^{i\frac{k_x a}{2}}$$

$$T_{13}' = -\frac{\sqrt{3}}{2}\, t_1(pd\sigma) e^{i\frac{k_y a}{2}}$$

$$T_{14}' = -\frac{1}{2} t_1(pd\sigma) e^{i\frac{k_y a}{2}}$$

$$T_{15}' = t_1(pd\pi) e^{i\frac{k_y a}{2}}$$

$$T_{16} = 0$$

$$T_{17} = 0$$

$$T_{18} = \frac{\sqrt{3}}{2}\, [\, -t(dd\sigma) + t(dd\delta)\,]\, (\cos k_x a - \cos k_y a)\,, \tag{10.17}$$

where

$$l_1 = \frac{0.5a}{\sqrt{(0.5a)^2 + (0.364c/2)^2}}$$

$$n_1 = \frac{0.364c/2}{\sqrt{(0.5a)^2 + (0.364c/2)^2}}$$

$$l_2 = \frac{0.5a}{\sqrt{2(0.5a)^2 + (0.272c/2)^2}}$$

$$n_2 = \frac{0.272c/2}{\sqrt{2(0.5a)^2 + (0.272c/2)^2}}\,. \tag{10.18}$$

Appendix C

By using the approximation that the effective-interaction term $\tilde{H}_{\text{int}}$ has non-vanishing matrix elements only for those between nearest neighbour atomic orbitals, we can represent the $\boldsymbol{k}$ dependence of the effective-interaction part of the Hamiltonian matrix, $\tilde{H}_{\text{int}}(\boldsymbol{k})$, as follows;

For on-site elements of $\tilde{H}^0_{AA}(\boldsymbol{k})$ and $\tilde{H}^0_{BB}(\boldsymbol{k})$,

$$E_1 = E_p^1$$
$$E_2 = E_p^2$$
$$E_3 = E_{dxy}$$
$$E_4 = E_{dxy}$$
$$E_5 = E_{dxy}$$
$$E_6 = E_{dx^2-y^2}$$
$$E_7 = E_{dx^2-y^2},$$

(10.19)

for the diagonal elements of $\tilde{H}^0_{AB}(\boldsymbol{k})$ and $\tilde{H}^0_{BA}(\boldsymbol{k})$,

$$E_1 = 0$$
$$E_2 = 0$$
$$E_3 = +2\, t(dd\pi)(\cos k_x a + \cos k_y a)$$
$$E_4 = +2\, t(dd\pi)\cos k_y a + 2\, t(dd\delta)\cos k_x a$$
$$E_5 = +2\, t(dd\pi)\cos k_x a + 2\, t(dd\delta)\cos k_y a$$
$$E_6 = +\left[\frac{3}{2} t(dd\sigma) + \frac{1}{2} t(dd\delta)\right](\cos k_x a + \cos k_y a)$$
$$E_7 = +\left[\frac{1}{2} t(dd\sigma) + \frac{3}{2} t(dd\delta)\right](\cos k_x a + \cos k_y a),$$

(10.20)

for the off-diagonal elements of $\tilde{H}^0_{AA}(\boldsymbol{k})$ and $\tilde{H}^0_{BB}(\boldsymbol{k})$

$$T_1 = \frac{1}{2}[\,t_1(pp\sigma) + t_1(pp\pi)\,](e^{i\frac{k_x a}{2}}e^{-i\frac{k_y a}{2}} + e^{-i\frac{k_x a}{2}}e^{i\frac{k_y a}{2}})$$
$$T_2 = -\frac{1}{2}[\,t_1(pp\sigma) - t_1(pp\pi)\,](e^{i\frac{k_x a}{2}}e^{-i\frac{k_y a}{2}} + e^{-i\frac{k_x a}{2}}e^{i\frac{k_y a}{2}})$$
$$T_3 = t_1(pp\pi)(e^{i\frac{k_x a}{2}}e^{-i\frac{k_y a}{2}} + e^{-i\frac{k_x a}{2}}e^{i\frac{k_y a}{2}})$$
$$T_4 = [\,l_1^2\, t_2(pp\sigma) + (1-l_1^2)\, t_2(pp\pi)\,]\,e^{-i\frac{k_x a}{2}}e^{i\,0.364 k_z c/2}$$
$$T_5 = -l_1 n_1\,[\,t_2(pp\sigma) - t_2(pp\pi)\,]\,e^{-i\frac{k_x a}{2}}e^{i\,0.364 k_z c/2}$$
$$T_6 = t_2(pp\pi)e^{-i\frac{k_x a}{2}}e^{i\,0.364 k_z c/2}$$
$$T_7 = [\,n_1^2\, t_2(pp\sigma) + (1-n_1^2)\, t_2(pp\pi)\,]\,e^{-i\frac{k_x a}{2}}e^{i\,0.364 k_z c/2}$$

$$T'_4 = [\, l_1^2\, t_2(pp\sigma) + (1 - l_1^2)\, t_2(pp\pi)\,]\, e^{-i\frac{k_y a}{2}} e^{i\, 0.364 k_z c/2}$$

$$T'_5 = -l_1 n_1\, [\, t_2(pp\sigma) - t_2(pp\pi)\,]\, e^{-i\frac{k_y a}{2}} e^{i\, 0.364 k_z c/2}$$

$$T'_6 = t_2(pp\pi) e^{-i\frac{k_y a}{2}} e^{i\, 0.364 k_z c/2}$$

$$T'_7 = [\, n_1^2\, t_2(pp\sigma) + (1 - n_1^2)\, t_2(pp\pi)\,]\, e^{-i\frac{k_y a}{2}} e^{i\, 0.364 k_z c/2}$$

$$T_8 = [\, l_2^2\, t_3(pp\sigma) + (1 - l_2^2)\, t_3(pp\pi)\,]\, (e^{i\frac{k_x a}{2}} e^{-i\frac{k_y a}{2}} + e^{-i\frac{k_x a}{2}} e^{i\frac{k_y a}{2}}) e^{i\, 0.272 k_z c/2}$$

$$T_9 = [\, n_2^2 t_3(pp\sigma) + (1 - n_2^2)\, t_3(pp\pi)\,]\, (e^{i\frac{k_x a}{2}} e^{-i\frac{k_y a}{2}} + e^{-i\frac{k_x a}{2}} e^{i\frac{k_y a}{2}}) e^{i\, 0.272 k_z c/2}$$

$$T_{10} = -l_2^2\, [\, t_3(pp\sigma) - t_3(pp\pi)\,]\, (e^{i\frac{k_x a}{2}} e^{-i\frac{k_y a}{2}} + e^{-i\frac{k_x a}{2}} e^{i\frac{k_y a}{2}}) e^{i\, 0.272 k_z c/2}$$

$$T_{11} = l_2 n_2\, [\, t_3(pp\sigma) - t_3(pp\pi)\,]\, (e^{i\frac{k_x a}{2}} e^{-i\frac{k_y a}{2}} - e^{-i\frac{k_x a}{2}} e^{i\frac{k_y a}{2}}) e^{i\, 0.272 k_z c/2}$$

$$T_{12} = l_2 n_2\, [t_3(pp\sigma) - t_3(pp\pi)\,]\, (-e^{i\frac{k_x a}{2}} e^{-i\frac{k_y a}{2}} + e^{-i\frac{k_x a}{2}} e^{i\frac{k_y a}{2}}) e^{i\, 0.272 k_z c/2}$$

$$T_{13} = -\frac{\sqrt{3}}{2}\, t_1(pd\sigma) e^{-i\frac{k_x a}{2}}$$

$$T_{14} = \frac{1}{2} t_1(pd\sigma) e^{-i\frac{k_x a}{2}}$$

$$T_{15} = -\, t_1(pd\pi) e^{-i\frac{k_x a}{2}}$$

$$T'_{13} = \frac{\sqrt{3}}{2}\, t_1(pd\sigma) e^{-i\frac{k_y a}{2}}$$

$$T'_{14} = \frac{1}{2} t_1(pd\sigma) e^{-i\frac{k_y a}{2}}$$

$$T'_{15} = -\, t_1(pd\pi) e^{-i\frac{k_y a}{2}}$$

$$T_{16} = t_2(pd\pi) e^{i\, 0.364 k_z c/2}$$

$$T_{17} = t_2(pd\sigma) e^{i\, 0.364 k_z c/2}$$

$$T_{18} = 0\,,$$

(10.21)

and for the off-diagonal elements of $\tilde{H}^0_{AB}(\boldsymbol{k})$ and $\tilde{H}^0_{BA}(\boldsymbol{k})$

$$T_1 = \frac{1}{2}\, [\, t_1(pp\sigma) + t_1(pp\pi)\,]\, (e^{i\frac{k_x a}{2}} e^{i\frac{k_y a}{2}} + e^{-i\frac{k_x a}{2}} e^{-i\frac{k_y a}{2}})$$

$$T_2 = \frac{1}{2}\, [\, t_1(pp\sigma) - t_1(pp\pi)\,]\, (e^{i\frac{k_x a}{2}} e^{i\frac{k_y a}{2}} + e^{-i\frac{k_x a}{2}} e^{-i\frac{k_y a}{2}})$$

$$T_3 = t_1(pp\pi) \cos\frac{k_x a}{2} (e^{i\frac{k_x a}{2}} e^{i\frac{k_y a}{2}} + e^{-i\frac{k_x a}{2}} e^{-i\frac{k_y a}{2}})$$

$$T_4 = [\, l_1^2\, t_2(pp\sigma) + (1 - l_1^2)\, t_2(pp\pi)\,]\, e^{i\frac{k_x a}{2}} e^{i\, 0.364 k_z c/2}$$

$$T_5 = l_1 n_1\, [\, t_2(pp\sigma) - t_2(pp\pi)\,]\, e^{i\frac{k_x a}{2}} e^{i\, 0.364 k_z c/2}$$

$$T_6 = t_2(pp\pi) e^{i\frac{k_x a}{2}} e^{i\, 0.364 k_z c/2}$$

$$T_7 = [\, n_1^2\, t_2(pp\sigma) + (1 - n_1^2)\, t_2(pp\pi)\,]\, e^{i\frac{k_x a}{2}} e^{i\, 0.364 k_z c/2}$$

$$T'_4 = [\, l_1^2\, t_2(pp\sigma) + (1 - l_1^2)\, t_2(pp\pi)\,]\, e^{i\frac{k_y a}{2}} e^{i\, 0.364 k_z c/2}$$

$$T'_5 = l_1 n_1\, [\, t_2(pp\sigma) - t_2(pp\pi)\,]\, e^{i\frac{k_y a}{2}} e^{i\, 0.364 k_z c/2}$$

10.5 Appendix   105

$$T'_6 = t_2(pp\pi)e^{i\frac{k_y a}{2}}e^{i\,0.364 k_z c/2}$$

$$T'_7 = [\,n_1^2\, t_2(pp\sigma) + (1-n_1^2)\, t_2(pp\pi)\,]\,e^{i\frac{k_y a}{2}}e^{i\,0.364 k_z c/2}$$

$$T_8 = [\,l_2^2\, t_3(pp\sigma) + (1-l_2^2)\, t_3(pp\pi)\,]\,(e^{i\frac{k_x a}{2}}e^{i\frac{k_y a}{2}} + e^{-i\frac{k_x a}{2}}e^{-i\frac{k_y a}{2}})e^{i\,0.364 k_z c/2}$$

$$T_9 = [\,n_2^2\, t_3(pp\sigma) + (1-n_2^2)\, t_3(pp\pi)\,]\,(e^{i\frac{k_x a}{2}}e^{i\frac{k_y a}{2}} + e^{-i\frac{k_x a}{2}}e^{-i\frac{k_y a}{2}})e^{i\,0.272 k_z c/2}$$

$$T_{10} = l_2^2\,[\,t_3(pp\sigma) - t_3(pp\pi)\,]\,(e^{i\frac{k_x a}{2}}e^{i\frac{k_y a}{2}} + e^{-i\frac{k_x a}{2}}e^{-i\frac{k_y a}{2}})e^{i\,0.272 k_z c/2}$$

$$T_{11} = l_2 n_2\,[\,t_3(pp\sigma) - t_3(pp\pi)\,]\,(e^{i\frac{k_x a}{2}}e^{i\frac{k_y a}{2}} - e^{-i\frac{k_x a}{2}}e^{-i\frac{k_y a}{2}})e^{i\,0.272 k_z c/2}$$

$$T_{12} = l_2 n_2\,[\,t_3(pp\sigma) - t_3(pp\pi)\,]\,(e^{i\frac{k_x a}{2}}e^{i\frac{k_y a}{2}} - e^{-i\frac{k_x a}{2}}e^{-i\frac{k_y a}{2}})e^{i\,0.272 k_z c/2}$$

$$T_{13} = \frac{\sqrt{3}}{2}\, t_1(pd\sigma)e^{i\frac{k_x a}{2}}$$

$$T_{14} = -\frac{1}{2}t_1(pd\sigma)e^{i\frac{k_x a}{2}}$$

$$T_{15} = t_1(pd\pi)e^{i\frac{k_x a}{2}}$$

$$T'_{13} = -\frac{\sqrt{3}}{2}\, t_1(pd\sigma)e^{i\frac{k_y a}{2}}$$

$$T'_{14} = -\frac{1}{2}t_1(pd\sigma)e^{i\frac{k_y a}{2}}$$

$$T'_{15} = t_1(pd\pi)e^{i\frac{k_y a}{2}}$$

$$T_{16} = 0$$

$$T_{17} = 0$$

$$T_{18} = \frac{\sqrt{3}}{2}\,[\,-t(dd\sigma) + t(dd\delta)\,]\,(\cos k_x a - \cos k_y a)\,,$$

(10.22)

where

$$l_1 = \frac{0.5a}{\sqrt{(0.5a)^2 + (0.364c/2)^2}}$$

$$n_1 = \frac{0.364c/2}{\sqrt{(0.5a)^2 + (0.364c/2)^2}}$$

$$l_2 = \frac{0.5a}{\sqrt{2(0.5a)^2 + (0.272c/2)^2}}$$

$$n_2 = \frac{0.272c/2}{\sqrt{2(0.5a)^2 + (0.272c/2)^2}}\,.$$

(10.23)

where a and b denote $2p_x$, $2p_y$ and $2p_z$ atomic orbitals for each of eight oxygen atoms and $3d_{yz}$, $3d_{xz}$, $3d_{xy}$, $3d_{x^2-y^2}$ and $3d_{z^2}$ atomic orbitals for each of two Cu atoms in the antiferromagnetic unit cell.

11 Calculated Results of Many-Electrons Band Structures and Fermi Surfaces

11.1 Introduction

In Chap. 10, for one of the approximation methods to solve the K–S Hamiltonian, we described a method of the mean-field approximation for treating a system of localized spins in the local antiferromagnetic (AF) order. By applying the mean-field approximation to the K–S Hamiltonian, the exchange interaction H_{ex} in the K–S Hamiltonian can be expressed in the form of an effective magnetic field acting on the spins of the hole-carriers in a carrier system.

As a result, the electronic structure of a hole-carrier system on the K–S model can be expressed in the form of a single-electron-type band structure in the presence of AF order in the localized spin system, where the exchange interaction between the spins of a hole-carrier and of a localized hole is included in a single-electron type energy band in the mean-field sense.

In this chapter we describe the results of the effective one-electron-type band structure and Fermi surface calculated by the method described in the previous chapter.

11.2 Calculated Band Structure Including the Exchange Interaction between the Spins of Hole-Carriers and Localized Holes

In a previous chapter, we showed that all the matrix elements in the 34×34 dimensional Hamiltonian matrix $\tilde{H}(\boldsymbol{k})$ are expressed as one-electron type quantities due to the mean-field approximation. By diagonalizing it, we obtained a one-electron type band structure including the many-electron effects such as the exchange interaction between the spins of a dopant hole and localized spin in the K–S Hamiltonian. In the effective one-electron type band structure thus obtained, the antibonding $\text{b}_{1\text{g}}^*$ orbitals which have a main character of Cu $d_{x^2-y^2}$ atomic orbital are separated from a hole-carrier system. These $\text{b}_{1\text{g}}^*$ orbitals are localized at Cu site by the strong U effect and the spins of localized holes in $\text{b}_{1\text{g}}^*$ orbitals are coupled antiferromagnetically by the effect of the superexchange interaction between the localized spins.

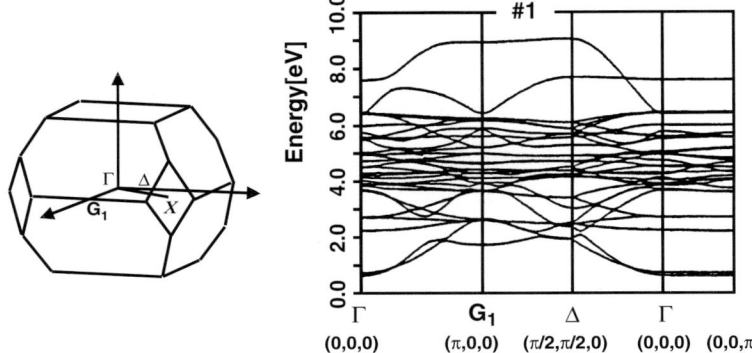

Fig. 11.1. The many-body-effect included band-structure for up-spin dopant holes, obtained by solving the effective one-electron-type 34×34 dimensional Hamiltonian matrix $\tilde{H}$ for an antiferromagnetic unit cell. The highest occupied band is marked by the #1 band. The Δ-point corresponds to $(\pi/2a, \pi/2a, 0)$, while the G_1-point to $(\pi/a, 0, 0)$. In this figure the Cu–O–Cu distance, a, is taken to be unity

In Fig. 11.1, the band structure for up-spin dopant holes calculated by Ushio and Kamimura for LSCO is shown for various values of wave-vector k and symmetry points in the antiferromagnetic (AF) Brillouin zone, where the AF Brillouin zone is also shown in the left hand side of the figure. The same band structure shape is also obtained for down-spin dopant holes. Here one should note that the energy in this figure is taken to be electron-energy but not hole-energy, and the Hubbard bands for localized b_{1g}^* holes do not appear in this figure.

In the undoped La_2CuO_4, all the bands except for the upper Hubbard b_{1g}^* band are occupied by electrons so that La_2CuO_4 is an insulator, consistent with experimental results. In this respect the present band structure is completely different from the ordinary energy band of the Fermi-liquid picture calculated by the local density approximation (LDA). The localized holes are accommodated in the upper Hubbard b_{1g}^* band, which consists mainly of Cu $d_{x^2-y^2}$ orbitals, leading to the formation of the AF spin ordering, while the dopant holes in the highest band in Fig. 11.1, marked by #1, have mainly O p_σ character. Thus the present theory shows that La_2CuO_4 is a mixed type of charge transfer insulator and Mott–Hubbard insulator, consistent with the experimental result [10, 163].

Now let us dope holes into this undoped La_2CuO_4. When Sr are doped, holes begin to occupy the top of the highest band in Fig. 11.1 marked by #1 (referred as the conduction band hereafter) at Δ point which corresponds to $(\pi/2a, \pi/2a, 0)$. At the onset concentration of superconductivity, x_c, the Fermi level E_F is located at the energy of $E = 9.04\,\mathrm{eV}$ just below the top of the #1 band at Δ, which is also a little higher than that of the G_1 point. Here the G_1 point in the AF Brillouin zone lies at $(\pi/a, 0, 0)$, and corresponds to a

11.2 Calculated Band Structure Including the Exchange Interaction

saddle point of the van Hove singularity as seen in Fig. 11.1. The characteristic feature of the #1 conduction band is the existence of *the flat band* along the line from G_1 to Δ. This feature is consistent with the ARPES data by Shen et al. [56] and Desseau et al. [164], who observed an extended region of flat band very near E_F around $\overline{M}$ point, which corresponds, in the present notation, to G_1 in the AF Brillouin zone, $(\pi/a, 0, 0)$.

The wave function of the conduction band for up-spin holes consists of a_{1g}^* orbitals at A-site and b_{1g} orbitals at B-site, as will be shown in Fig. 11.9 in Sect 11.4. Besides the localized b_{1g}^* holes in the upper Hubbard bands, a_{1g}^* orbitals at A-site and b_{1g} orbitals at B-site form the $^3B_{1g}$ multiplet and $^1A_{1g}$ multiplet, respectively. Thus the present calculated results realize the electronic structure of the K–S model, where the hole-carriers take $^3B_{1g}$ and $^1A_{1g}$ alternately in the spin-correlated region of the local AF order.

In the present calculation we have assumed the long range Néel order, while the results of neutron inelastic scattering experiments [145] suggest that the localized spins in a two-dimensional (2D) CuO_2 plane are fluctuating and there is no long range Néel order in the superconducting regime, although the local AF order has been observed. Thus it is necessary to discuss how the spin fluctuation of localized spins in the 2D Heisenberg AF order affects the electronic and magnetic properties of LSCO. Let λ_s be a characteristic length of the spin-correlated region, in which the coherent motion of a dopant hole is retained due to the existence of the local AF order.

In this spin-correlated region, the frustrated spins on its boundary change their directions by the fluctuation effect in the 2D Heisenberg AF spin system during the time of τ_s defined by $\tau_s \equiv \hbar/J$, with J being the superexchange interaction (~ 0.1 eV) [146]. In this case the hole-carriers at the Fermi level may move coherently much longer than the observed spin-correlation length, when the traveling time of a hole-carrier at the Fermi level over an area of the spin-correlation length, which is given by $\tau_F \equiv \lambda_s/v_F$, is longer than τ_s, where v_F is the Fermi velocity of a hole-carrier at the Fermi level. This is the case for underdoped LSCO, because τ_s is of the order of 10^{-15} sec while τ_F is of the order of 10^{-14} sec for the underdoped region of $x = 0.10$ to $x = 0.15$ in LSCO. In this way the region of a metallic state becomes much wider than the spin-correlation length so as to reduce the increase of the kinetic energy due to the confinement of hole-carriers in the spin-correlated region. A behaviour of coherent motion of a hole-carrier across the boundary of a spin-correlated region thus described is schematically shown in Fig. 11.2. The length of a wider metallic region is denoted by ℓ_0. This ℓ_0 is much wider than λ_s. The electronic, thermal and magnetic properties of cuprates are determined by the hole-carriers in a metallic region of ℓ_0.

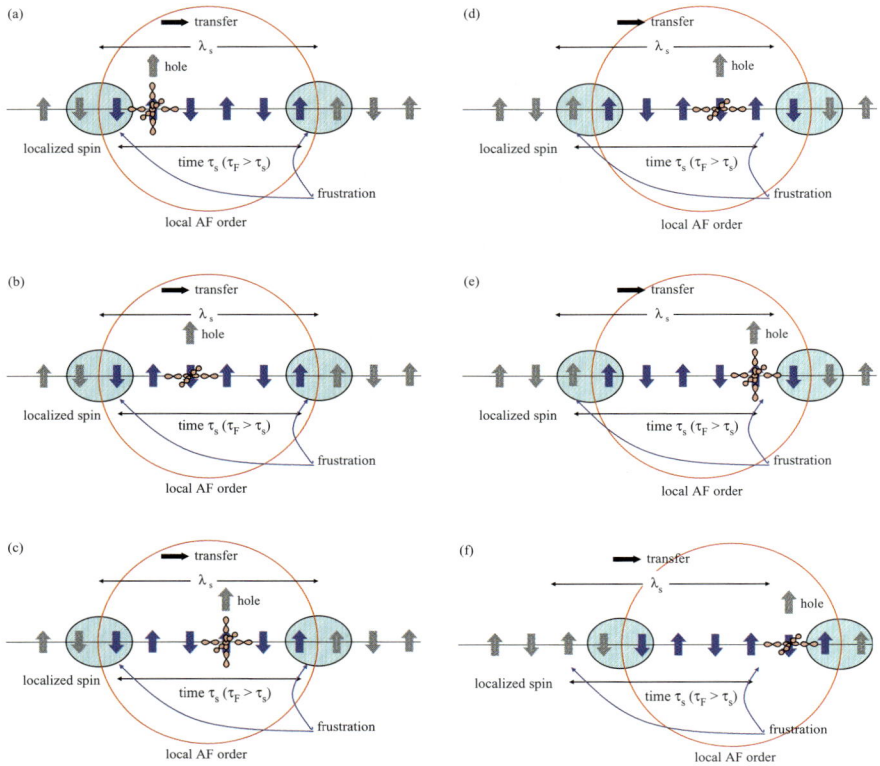

Fig. 11.2. Schematic picture of a hole-carrier motion across the boundary of a spin-correlated region λ_s, by the 2D spin-fluctuation effect. By choosing a system of 11 localized spins along one direction, this picture shows schematically why a metallic region is wider than that of the local AF region

11.3 Calculated Fermi Surface and Comparison with Experiments

Now we construct the Fermi surfaces (FS), based on the calculated conduction band shown in Fig. 11.1. This Fermi surface is also completely different from that of an ordinary Fermi liquid picture calculated by the local density approximation (LDA), as already pointed out in a previous section, because the conduction band in the present result is fully occupied by electrons in the undoped case while the LDA band always yields a metallic state. Further, a carrier system with up or down spin has a respective Fermi surface, although their shape and their position in **k**-space are the same. The Fermi surface is constructed by connecting the points in the **k**-space at which the Fermi distribution function shows discontinuity. In Fig. 11.3 the Fermi surface structure thus obtained for $x = 0.15$ is shown as an example, where one Fermi surface

11.3 Calculated Fermi Surface and Comparison with Experiments

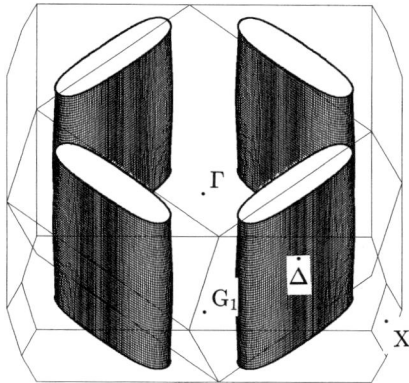

Fig. 11.3. The Fermi surface for $x = 0.15$ calculated for the #1 band. Here two kinds of Brillouin zones are also shown. One at the outermost part is the ordinary Brillouin zone corresponding to an ordinary unit cell consisting of a single CuO_6 octahedron and the inner part is the folded Brillouin zone for the AF unit cell in LSCO. Here the k_x axis is taken along $\overline{\Gamma G_1}$, corresponding to the x-axis (the Cu–O–Cu direction) in a real space

consists of two pairs of extremely flat tubes. Two Fermi surfaces of each pair face each other along bisectors between k_x- and k_y-axes. These Fermi surfaces in each pair are separated by the reciprocal lattice vectors in the AF Brillouin zone; $\boldsymbol{Q}_{AF1} = (\pi/a, \pi/a, 0)$ or $\boldsymbol{Q}_{AF2} = (-\pi/a, \pi/a, 0)$, i.e., AF reciprocal unit vectors. The cross-section of each Fermi surface is very small as seen in Fig. 11.3. This unique feature of the Fermi surface structure is consistent with recent experimental results of the angle-resolved photoemission (ARPES) by various experimental groups.

In order to compare with the experimental results of ARPES for LSCO, we have calculated the Fermi surface of in $La_{2-x}Sr_xCuO_4$ for the following hole-concentrations $x = 0.025$, $x = 0.05$, $x = 0.075$, $x = 0.1$, $x = 0.125$, and $x = 0.15$. These Fermi surfaces are projected on the $k_x - k_y$ plane of the two-dimensional AF Brillouin zone. These are shown in Figs. 11.4(a), (b), (c), (d), (e) and (f). As seen in these figures, the calculated Fermi surfaces are small. However, because of the finite life time of carriers due to the finite size of a metallic region in which the K–S model can be applied, a part of Fermi surfaces in the second AF Brillouin zone is not observed definitely. Recently the Fermi arcs in the first AF Brillouin zone are observed clearly by the ARPES experiment for $La_{2-x}Sr_xCuO_4$ by Yoshida et al. [23, 24, 25]. For example, the experimental results of Fermi arcs for $La_{2-x}Sr_xCuO_4$ from $x = 0.03$ to 0.15 are shown in Fig. 11.5, where points in the central parts of bow-shape strips in the figure correspond to the Fermi arcs.

In Fig. 11.6, the calculated results shown in Figs. 11.4(a), (b), (c), (d), (e) and (f) are superposed on the experimental results by Yoshida et al. [25].

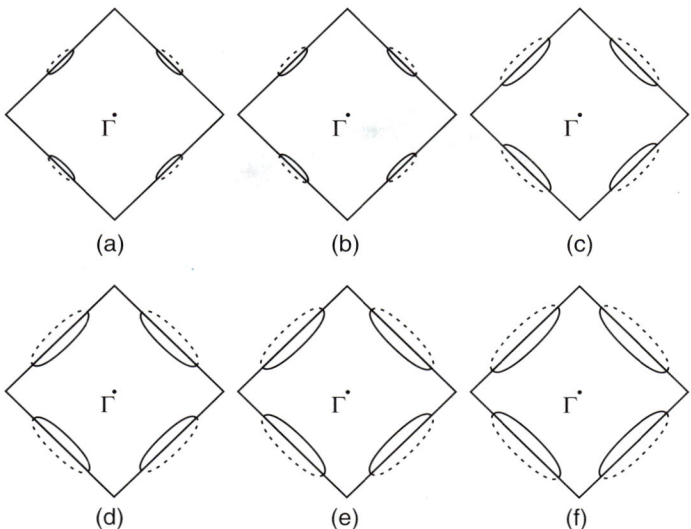

Fig. 11.4. The calculated Fermi surface (FS) of LSCO for (**a**) $x = 0.025$, (**b**) $x = 0.05$, (**c**) $x = 0.075$, (**d**) $x = 0.1$, (**e**) $x = 0.125$, and (**f**) $x = 0.15$ based on K–S model, drawn in the two dimensional AF Brillouin zone for the convenience of comparison with the results observed by ARPES. *Inner* edges of FS which correspond to the so-called Fermi arcs are shown by *solid lines* while the *outer* edges are represented by *dotted lines* due to vagueness from the lifetime broadening

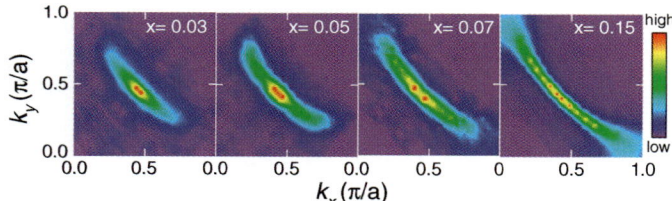

Fig. 11.5. Doping dependence of Fermi surfaces of $La_{2-x}Sr_xCuO_4$ from $x = 0.03$ to 0.15, observed in ARPES by T. Yoshida et al. after [25] and [23]. Points in the central parts of bow-shape strips represent Fermi arcs

As seen in Fig. 11.6, the agreement between the calculated results and the observed ones is surprisingly good. Thus we can say that recent ARPES experiments by Yoshida et al. provide clear experimental evidence for the K–S model.

As regards other cuprates such as superconducting $Bi_2Sr_{0.97}Pr_{0.03}CuO_{6+\delta}$ (Bi2201) compounds which includes a single CuO_2 layer in a unit cell like LSCO [165], the observed Fermi surface structure is very similar to the present theoretical results. Fermi surface structures for $Bi_2Sr_2CaCu_2O_{8+\delta}$ (Bi2212) determined by angle-resolved photoemission [56, 160, 164, 166] are also very

11.3 Calculated Fermi Surface and Comparison with Experiments

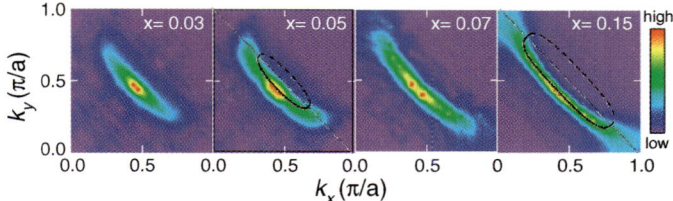

Fig. 11.6. Observed doping dependence of Fermi surfaces of $La_{2-x}Sr_xCuO_4$ from $x = 0.03$ to 0.15, observed in ARPES by T. Yoshida et al. The calculated results based on the K–S model for $x = 0.05$ and $x = 0.15$ shown by *thin* curves are superimposed on the experimental results obtained by Yoshida et al.

similar to the present result, although the Fermi surface structure for Bi2212 is more complicated due to the existence of two CuO_2 layers in a unit cell. Shen et al. [56] and Desseau et al. [164] mapped out the near-E_F electronic structure and Fermi surface of Bi2212 by angle-resolved photoemission. They observed an extended region of the flat CuO_2 derived bands very near E_F around $\overline{M}$ point, which is the G_1 point in our notation, and also the strong tendency of the nesting of the Fermi surface along the nesting vectors $\boldsymbol{Q}_1$ or $\boldsymbol{Q}_2$. Further, Aebi et al. [160, 166] found a $c(2 \times 2)$ superstructure on the Fermi surface, suggesting the short range antiferromagnetic correlation, consistent with the prediction by Ushio and Kamimura. Further, Marshall et al. [43] also observed a small Fermi surface structure for the underdoped Dy concentration of $Bi_2Sr_2Ca_{1-x}Dy_xCu_2O_{8+\delta}$ with $T_c = 65$ K, consistent with the prediction of a small Fermi surface by Kamimura and Ushio.

Although the present calculation is based on a periodic system with the antiferromagnetic order, in a real system the spin correlation length is finite so that the appearance of the small Fermi surface structure has a finite lifetime. As a result, various phenomena based on the present small Fermi surface structure are expected to have lifetime broadening effects. For example, the outer edge of each section in the Fermi surface structure shown by dotted lines in Fig. 11.4(a)–(e) is not sharp compared with its inner edge in the first AF Brillouin zone due to the above lifetime broadening effect, so that it might be very difficult to see both edges of each section in the Fermi surface clearly in the angle-resolved photoemission experiments. This is one of the reasons why the angle-resolved photoemission experiments can not clearly determine whether the Fermi surfaces are large or small. When the spin-correlation length becomes smaller with increasing hole-concentration, the regions of antiferromagnetic ordering become comparable to the mean-free path of carriers from the over-doped to the well over-doped region. Thus the K–S model does not hold in these hole concentration region. In particular, the superexchange interaction between localized spin via intervening O^{2-} ions is destroyed when the hole-concentration increases in the overdoped region. As a result small Fermi surfaces change into a large Fermi surface. This may explain the "cross-over phenomena" observed in various normal state transport properties. For example, the spin susceptibility change from a 2D AF characteristic to a Pauli-like behaviour, and the high-energy pseudogap appears.

In connection with the appearance of the small Fermi surface one might question whether the Fermi surface should include the contribution from the localized spin or not in connection with Luttinger's theorem [167]. The present small Fermi surface does not include the contribution from localized spins. However, this does not contradict the Luttinger theorem because the antiferromagnetic order coexists locally in the present case.

Finally a brief remark is made on the origin of the incommensurate peak observed in the inelastic neutron scattering experiment [168, 169]. A possible explanation for the origin of the incommensurate peaks is given by the unique shape of the Fermi surface structure based on the K–S model. As we have already pointed out, there is a possibility of nesting between Fermi surfaces with different spins along the nesting wave vectors of $\boldsymbol{Q}_\gamma$ with $\gamma = 1$ and 2, which are deviated from an AF reciprocal unit vector $\boldsymbol{Q}_{\mathrm{AF1}} = (\pi, \pi, 0)$ by $(-\delta\pi, 0, 0)$ and $(0, -\delta\pi, 0)$, or along the nesting wave vectors of $\boldsymbol{Q}_\gamma$ with $\gamma = 3$ and 4, which are deviated from another AF reciprocal unit vector $\boldsymbol{Q}_{\mathrm{AF2}} = (-\pi, \pi, 0)$ by $(\delta\pi, 0, 0)$ and $(0, -\delta\pi, 0)$. The latter case is shown in Fig. 11.7. This nesting may be related to the appearance of incommensurate peaks in the spin excitation spectra of LSCO observed by neutron diffraction experiments [168, 169]. The incommensurability δ observed in neutron diffraction experiments for LSCO is compared with the calculated values of δ which are determined from the position of the nesting wave vector $\boldsymbol{Q}_\gamma$ for the calculated Fermi surfaces. The calculated results are plotted with open circles in Fig. 11.8, for various Sr concentrations. It shows non-linearity, consistent with the experimental results by Endoh et al. [170] and Thurston et al. [171].

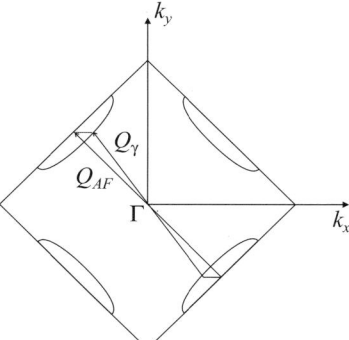

Fig. 11.7. The Fermi surface in the k_x-k_y plane for $x = 0.1$ in the AF Brillouin zone of LSCO, and schematic view of the nesting vectors $\boldsymbol{Q}_\gamma$ with $\gamma = 3$ or 4 and the AF reciprocal unit vector $\boldsymbol{Q}_{\mathrm{AF2}} = (-\pi, \pi, 0)$

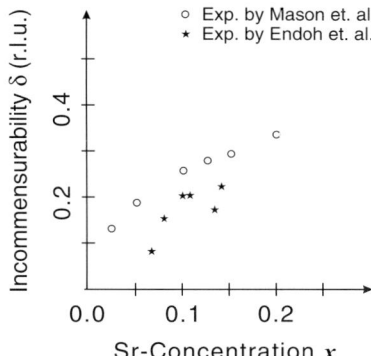

Fig. 11.8. Incommensurability δ in reciprocal lattice unit (r.l.u.). The *open circles* are the calculated results, while *daggers* and *stars* represent experimental results by Mason et al. [168] and Endoh et al. [170] respectively

As regards small Fermi surfaces, we should mention that Wen and Lee also obtained theoretically small Fermi pockets at low doping, which continuously evolve into a large Fermi surface at high doping concentrations [172].

11.4 Wavefunctions of a Hole-Carrier with Particular k Vectors and the Tight Binding (TB) Functional Form of the #1 Conduction Band

In Fig. 11.9 the wave functions of an up-spin carrier in the antiferromagnetic (AF) unit cell are shown for Δ and G_1 points in the AF Brillouin zone, where the right hand side of the figure corresponds to a CuO_6 cluster with localized up-spin (A-site) while the left hand side corresponds to a CuO_6 cluster with localized down-spin (B-site) in the AF unit cell. In this figure a main orbital component at each site is shown for A and B sites.

The calculated values for the mixing ratio of the in-plane Op_σ, apical Op_z, Cu $d_{x^2-y^2}$ and d_{z^2} orbitals in the wave function for five k values along the line from G_1 to Δ in the AF Brillouin zone are shown in Table 11.1, where the probability of finding a hole in each atomic orbital is shown. One can see from Fig. 11.9 and Table 11.1 that, in a certain concentration below the onset of superconductivity in which the Fermi level E_F crosses the #1 band at k_2 in Table 11.1, the holes with up-spin are accommodated in b_{1g} orbital constructed mainly from the oxygen p_σ orbitals in a CuO_2 plane, consistent with the result of the cluster calculation by Kamimura and Eto[104], while in the superconducting concentration region, which corresponds to k_3 and k_4 in Table 11.1, the holes move coherently from a_{1g}^* orbital at the A-site to b_{1g} orbital at the B-site.

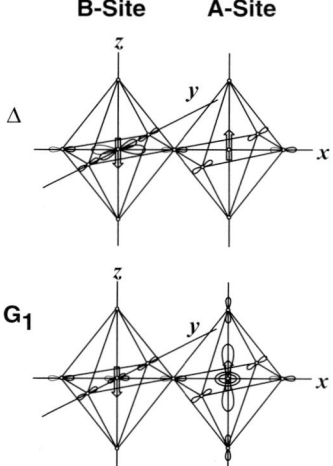

Fig. 11.9. The wave functions at Δ and G_1 points. Here, the *right hand side* of the figure corresponds to a CuO_6 cluster with localized up-spin (A-site) and the *left hand side* to a CuO_6 cluster with localized down-spin (B-site)

Table 11.1. The mixing ratio of the in-plane Op_σ, apical Op_z, Cu $d_{x^2-y^2}$ and Cu d_{z^2} orbitals in the wave functions for five k values in the AF Brillouin zone along the line G_1 to Δ

k	(k_x, k_y, k_z)	in-plane Op_σ	apical Op_z	Cu $d_{x^2-y^2}$	Cu d_{z^2}
k_1	$(\frac{1}{2}\pi, \frac{1}{2}\pi, 0)$	0.61	0.0	0.39	0.0
k_2	$(\frac{3}{8}\pi, \frac{5}{8}\pi, 0)$	0.58	0.003	0.37	0.05
k_3	$(\frac{1}{4}\pi, \frac{3}{4}\pi, 0)$	0.49	0.01	0.33	0.17
k_4	$(\frac{1}{8}\pi, \frac{7}{8}\pi, 0)$	0.41	0.02	0.29	0.28
k_5	$(0, \pi, 0)$	0.38	0.02	0.28	0.32

The calculated result in Table 11.1 shows that the mixing ratio of the $^1A_{1g}$ state to the $^3B_{1g}$ state at the k_2 value of $(3\pi/8, 5\pi/8, 0)$ in the underdoped regime of LSCO is 7 to 1. Thus, although the alternating appearance of the Zhang–Rice singlet $^1A_{1g}$ and the Hund's coupling triplet $^3B_{1g}$ is a characteristic feature of the K–S model in the underdoped superconducting regime of LSCO, the result of Table 11.1 indicates that the weight of the $^1A_{1g}$ state in the many-electron wave function of the K–S model is about seven times larger than that of the $^3B_{1g}$. This result is consistent with the experimental one obtained the polarized X-ray absorption spectra (XAS) by C. T. Chen et al. [93].

Reflecting the alternate appearance of a$^*_{1g}$ and b$_{1g}$ orbitals among A and B sites, the top of the conduction band (#1 band) in Fig. 11.1 appears at the Δ point in the AF Brillouin zone, where the Δ point corresponds to ($\pi/2a$, $\pi/2a$, 0). The conduction band is approximately expressed in the following tight binding (TB) form;

$$\begin{aligned}
\epsilon_{\boldsymbol{k}} = \ & A \, [\cos(ak_x + ak_y) + \cos(-ak_x + ak_y)] \\
& + B \, \cos(ak_x + ak_y)\cos(-ak_x + ak_y) \\
& + C \, \cos(ak_x + ak_y)\cos(-ak_x + ak_y) \cos\frac{a}{2}k_x \cos\frac{a}{2}k_y \cos\frac{c}{2}k_z \\
& + D \, [\cos(ak_x + ak_y) + \cos(-ak_x + ak_y)] \cos\frac{a}{2}k_x \cos\frac{a}{2}k_y \cos\frac{c}{2}k_z \\
& + E_0 \, .
\end{aligned} \quad (11.1)$$

Here a and c are the lattice constants of the tetragonal unit cell, where $a = 3.78$ Å and $c = 13.25$ Å. The values of coefficient A to E_0 in (11.1) are determined so as to reproduce the numerically calculated conduction band by Ushio and Kamimura [112]. The values of A to E_0 thus determined are $A = -0.3311$ eV, $B = -0.3936$ eV, $C = -0.0006$ eV, $D = -0.0047$ eV and $E_0 = 8.647$ eV. The lower energy region of the conduction band below 8.0 eV in Fig. 11.1 does not fit well to the one calculated numerically. However, this disagreement does not influence the calculated results in the underdoped region because only the higher energy region of the conduction band above $E \geq 8.9$ eV contributes to the electronic structure for the hole concentration region of $x \leq 0.4$.

11.5 Calculated Density of States

We have also calculated the density of states of the conduction band (#1 band) in LSCO. The calculated density of states is shown as a function of energy in Fig. 11.10, where the origin of the energy is taken at the top of the conduction band (# 1 band) at the Δ point. The density of states for the conduction band has a sharp peak at E_F corresponding to $x \sim 0.3$ in La$_{2-x}$Sr$_x$CuO$_4$. The appearance of this sharp peak is due to a modified type of a saddle point singularity at G$_1$ point, as described below. The energy of the conduction band near the G$_1$ point increases towards the Δ point (along the direction of a point $(\pm\frac{\pi}{a}, \pm\frac{\pi}{a}, 0)$), while it decreases towards the Γ point (along the direction of a point $(\pm\frac{\pi}{a}, 0, 0)$ or a point $(0, \pm\frac{\pi}{a}, 0)$). Thus the G$_1$ point corresponds to the van-Hove singularity of the saddle-point type.

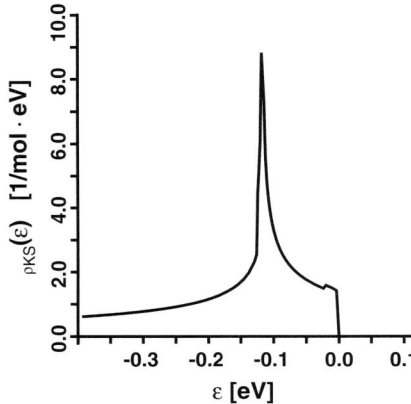

Fig. 11.10. The density of states of LSCO as a function of energy. The *solid lines* are those calculated from the #1 band in the renormalized band structure [113, 173]. The energy is measured from the top of the band. Holes enter from the top of the band at energy = 0

11.6 Remarks on the Simple Folding of the Fermi Surface into the AF Brillouin Zone

In this section we would like to remark that the appearance of the center of the Fermi surfaces at the Δ point is due to the alternant appearance of $^1A_{1g}$ and $^3B_{1g}$ multiplets among A and B sites, but not the result of a simple folding of the ordinary Brillouin zone into the AF Brillouin zone. Thus the presence of Fermi surfaces in the underdoped region is a characteristic feature of the K–S model with a two-component scenario. In this respect let us assume now a simple folding of a b_{1g} band in the presence of the AF order. Then the dopant holes are accommodated from the top of the upper branch of the folded b_{1g} band at Γ-point, which corresponds to the k value of $(0, 0, 0)$, since the undoped La_2CuO_4 is the mixture of Mott–Hubbard type- and of charge-transfer type-insulator, and both the upper and the lower branches of the folded b_{1g} bands are fully occupied by electrons in the undoped case. Therefore the center of the Fermi surface is at Γ-point in this case. On the other hand, according to the present calculation, the b_{1g} and a_{1g}^* bands split into four bands in the presence of AF order. The upper two bands among the four bands correspond to a character consisting of a_{1g}^* orbitals at A-site and b_{1g} orbitals at B-site while the character of lower two bands consist of a_{1g}^* orbitals at B-site and b_{1g} orbitals at A-site. Since both the b_{1g} and the a_{1g}^* bands are fully occupied by electrons in the undoped La_2CuO_4, the dopant holes are accommodated from the top of the conduction band (#1 band), and thus the character of the highest band is not pure b_{1g} orbitals, but the mixture of two kinds of orbitals, a_{1g}^* and b_{1g}. Therefore, the result of the present calculation that the Δ-point is the top of this highest band is

11.6 Remarks on the Simple Folding of the Fermi Surface

not obtained by a simple folding of an energy band due to the presence of the AF order. Thus, in order to obtain the present Fermi surface structure, it is essential to take account of the alternating appearance of b_{1g} and a_{1g}^* orbitals for a dopant hole in addition to b_{1g}^* orbital for a localized spin. Thus the present Fermi structure of the underdoped region shown in Fig. 11.3 and 11.4 is characteristic of the K–S model.

12 Normal State Properties of $La_{2-x}Sr_xCuO_4$

12.1 Introduction

In Chap. 10 we described the mean-field approximation as one of approximate methods to solve the K–S Hamiltonian of (11.1). Perhaps the most important consequence of this approximation is that, having dealt with the strong exchange interactions between the spins of hole-carriers and the localized spins in the fourth term of the K–S Hamiltonian in the mean field approximation, the transport, thermal and paramagnetic properties of the underdoped cuprates can be calculated within the framework similar to a single particle band structure. In this case, an assumption of a large size of an AF lattice is made, but the exchange interactions between the spins of dopant and localized holes are included in the form of an effective magnetic field in the present first principles calculations of the many-body energy band structures for a carrier system, as we showed in Chaps. 10 and 11. Hence, as we described in Chap. 11, the hole-carriers are to be found in the many-body energy bands consisting mainly of the characters of copper $d_{z^2}(a_{1g}^*)$ and oxygen $p_\sigma(b_{1g})$ orbitals, which are full in the un-doped state, leading to the Mott–Hubbard insulator, and the dopant holes can move relatively freely in the CuO_2 plane by taking the character of a Zhang–Rice singlet and Hund's coupling triplet alternately between neighbouring sites in the presence of the AF order due to the localized spins. The carrier is thus mobile while the underlying AF ordering is preserved, resulting in a metallic state with itinerant holes.

In this chapter we will calculate various normal state properties such as the electrical resistivity, Hall effect, electronic entropy, etc., by using the many-body included energy bands, the density of states and Fermi surfaces obtained in the mean-field approximation for the K–S Hamiltonian. We will show in this chapter that observed anomalous behaviours of various normal state properties in the underdoped region of cuprates can be explained successfully by the K–S model without introducing disposal parameters. Finally we will also discuss how the observed "high-energy-pseudogap" can be explained on the basis of the K–S model.

12.2 Resistivity

We begin with discussing the anomalous behaviour of electrical resistivity of underdoped cuprates by choosing LSCO as an example of the present study. As regards the electrical resistivity of $La_{2-x}Sr_xCuO_4$ (sometimes abbreviated as LSCO), Nakamura and Uchida [174] measured the temperature dependence of the in-plane and out-of-plane resistivity with regard to the CuO_2 plane of a single crystal of LSCO. Their result is shown in Fig. 12.1, where the upper and lower panels show, respectively, the in-plane- and out-of-plane-resistivity of LSCO for various hole concentrations (x). These experimental results show a remarkable difference in the temperature dependence of the underdoped regime between the in-plane-resistivity ρ_{ab} and out-of-plane resistivity ρ_c. In other words, (1) as regards ρ_{ab}, it shows a linear temperature-dependent metallic conductivity in a wide temperature range from high temperatures above room temperature down to T_c, as already shown by Takagi and his coworkers for epitaxial films [10, 175], while (2) ρ_c is two orders of magnitude larger in its values than those of ρ_{ab} for every x and it shows a striking feature of non-metallic behaviour in its temperature dependence for underdoped hole concentrations. Those anomalous behaviours of ρ_{ab} and ρ_c mentioned above can be explained naturally by the K–S model from a qualitative point of view, as seen below: Since the Fermi surfaces in the K–S model for the underdoped regime are small, the electron–phonon scatterings with small momentum transfer are possible. As a result the resistivity

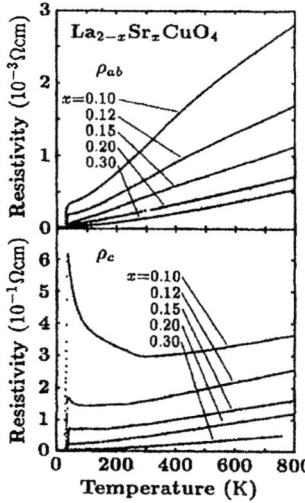

Fig. 12.1. Temperature dependence of the in-plane (*upper panel*) and out-of-plane (*lower panel*) resistivity for single crystals with various concentration of LSCO (After Nakamura and Uchida [174])

depends on temperature linearly down to very low temperatures near T_c. Thus the linear temperature dependence of ρ_{ab} can be explained by the ordinary mechanism of electron–phonon scattering for resistivity, based on the K–S model. On the other hand, one can explain the non-metallic behaviour of ρ_c in the underdoped regime by the characteristic feature of the K–S model that the metallic regions in a CuO_2 plane are distributed inhomogeneously. In other words, since the inhomogeneous distribution of metallic regions over the CuO_2 planes are different for every CuO_2 plane, the magnitudes of the transfer interactions of a hole-carrier between inter-planes in the underdoped region is random and small, depending on the way of distribution of metallic regions on two neighbouring CuO_2 planes. As a result, the conduction mechanism along the c-axis is due to hopping rather than a coherent transfer for the underdoped region. In this context we predict that the temperature- and concentration-dependence of the c-axis conduction in a cuprate has a semiconducting feature. Now we will calculate the temperature dependence of the in-plane resistivity along the ab plane ρ_{ab}. The Fermi surface structure of the K–S model in the k_x–k_y plane of the AF Brillouin zone is very similar to that of the higher-stage graphite intercalation compounds (GICs), in which the Fermi surface consists of four pockets of small area. In calculating the in-plane resistivity of higher-stage GICs, Inoshita and Kamimura found that the resistivity is proportional to T in the low temperature region due to the intra-pocket scattering and to T^2 in the high temperature region due to the inter-pocket scattering [176]. Following their method we will calculate the temperature dependence of the in-plane resistivity by adopting a simplified model of small Fermi surfaces(SF) shown in Fig. 11.3 of Chap. 11. For this purpose the phonon-limited resistivity in LSCO is calculated from the well-known variational expression for the resistivity of metals [177]. The resistivity formula due to collisions of the hole-carriers with lattice phonons is given below [176, 177],

$$\rho(T) = \frac{A\pi}{2e^2 k_B T} \int\int \omega_q |g_{k,K}|^2 \cdot \left[\cosh\left(\frac{\hbar\omega_q}{k_B T}\right) - 1\right]^{-1}$$
$$\times [(\boldsymbol{v_k} - \boldsymbol{v_K}) \cdot \boldsymbol{u}]^2 \frac{dS_k}{v_k}\frac{dS_K}{v_K} , \qquad (12.1)$$

with

$$A = \left[\int\int (\boldsymbol{v_k}\cdot\boldsymbol{u})(\boldsymbol{v_K}\cdot\boldsymbol{u})(\boldsymbol{v_k}\cdot\boldsymbol{v_K})\frac{dS_k}{v_k}\frac{dS_K}{v_K}\right]^{-1} , \qquad (12.2)$$

where $g_{k,K}$ is the electron–phonon matrix element between the states of the wave vectors $\boldsymbol{k}$ and $\boldsymbol{K}$ of an electron which interacts with a phonon of the wave vector $\boldsymbol{q}$ and the frequency ω_q with $\boldsymbol{q} = \boldsymbol{K} - \boldsymbol{k}$, $\boldsymbol{v_k}$ the group velocity of an electron in the state $\boldsymbol{k}$, which is given by $\boldsymbol{v_k} = \partial E_k/\partial \boldsymbol{k}$, $\hbar\omega_q$ the phonon energy, $\boldsymbol{u}$ the unit vector in the direction of the external electric field which is parallel to the x-axis, and $\int dS_k$ denotes an integration over the Fermi surface. Since the Fermi surface section in the k_x–k_y plane is small

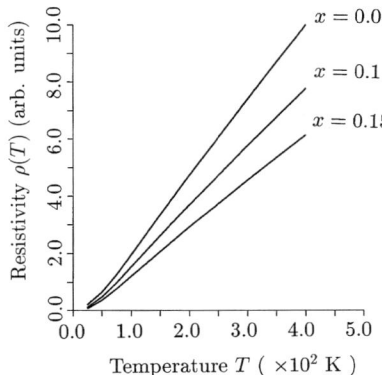

Fig. 12.2. The calculated temperature dependence of the resistivity in the ab plane, ρ_{ab}, of LSCO for various hole concentrations

in the K–S model, the phonons of small wave vectors are involved in the mechanism of causing resistivity. Thus $g_{\bm{k},\bm{K}}$ is expressed by the following form: $|g_{\bm{k},\bm{K}}|^2 = (N/2M\omega_{\bm{q}}) \cdot (\bm{C} \cdot \bm{q})^2$, with $\bm{C}$ being a coupling constant whose dimension is energy. Further, the phonon dispersion along the c-axis is small, so that one can express the phonon dispersion in a two-dimensional $\bm{q}$ space as $\hbar\omega_{\bm{q}} = v_s \cdot q_\perp = v_s \cdot \sqrt{q_x^2 + q_y^2}$, with the sound velocity v_s, where v_s is 5×10^5 cm/s for LSCO [178].

The calculated results of resistivity in the ab plane of LSCO are shown in Fig. 12.2 as a function of temperature T for $x = 0.05, 0.1$ and 0.15 in $\mathrm{La}_{2-x}\mathrm{Sr}_x\mathrm{CuO}_4$. Because the resistivity in the underdoped region of LSCO is governed by the electron–phonon scattering with small momentum transfer inside the same Fermi surface of small area, a linear temperature-dependence of the resistivity appears even in a low temperature region, such as below 150 K, consistent with the observed resistivity in normal state of $\mathrm{La}_{2-x}\mathrm{Sr}_x\mathrm{CuO}_4$ by Nakamura and Uchida [174] and Takagi et al. [175]. The calculated concentration dependence of ρ_{ab} is also consistent with experimental results by Nakamura and Uchida [174] and Takagi et al. [175]. For the values of x above 0.18, the observed temperature dependence in resistivity deviates upward from the linear dependence in a low temperature region. Since the K–S model does not hold in the well-overdoped region and Fermi surfaces change from small ones to a larger one, the above mentioned deviation from T-linear resistivity in the low-temperature region may be due to this effect.

According to Inoshita and Kamimura, the electron–phonon scattering between the neighbouring FS pockets becomes dominant above 150 K, and this gives rise to T^2 temperature dependence in ρ_{ab}. In the higher temperature region, however, the electron–phonon scattering within a large FS will contribute to the T-linear temperature dependence to ρ_{ab}. When temperature increases, the slope of T-linear dependence becomes different, from that in the low-T region, because the K–S model will not hold in higher temperature region so that small Fermi surfaces will change to a large FS, as will be seen later in Sects. 12.5 and 12.6.

12.3 Hall Effect

The observed Hall coefficient of LSCO, R_H, has also shown an anomalous behaviour. The Hall data for LSCO show a drop in R_H by two orders of magnitude as x increases from $x = 0.1$ to 0.3. Then a sign change of R_H from a hole-like to an electron-like character occurs at around $x = 0.3$ [10, 179]. In this section we will show that these anomalous behaviours of the Hall coefficient can be explained by the K–S model without introducing any adjustable parameters.

For this purpose we use the formula derived by Schimizu and Kamimura [180], by substituting $-\frac{\partial f}{\partial E_k}$ for the δ-function in their formula. The formula for the Hall coefficient R_H thus obtained is given as follows:

$$R_H = \frac{4\pi^3}{ec} \frac{\int_{BZ} d\boldsymbol{k} \frac{\partial E_{\boldsymbol{k}}}{\partial k_x} \left[\frac{\partial E_{\boldsymbol{k}}}{\partial k_x} \frac{\partial^2 E_{\boldsymbol{k}}}{\partial k_y^2} - \frac{\partial E_{\boldsymbol{k}}}{\partial k_y} \frac{\partial^2 E_{\boldsymbol{k}}}{\partial k_x \partial k_y} \right] \left(-\frac{\partial f}{\partial E_{\boldsymbol{k}}} \right)}{\left[\int_{BZ} d\boldsymbol{k} \left(\frac{\partial E_{\boldsymbol{k}}}{\partial k_x} \right)^2 \left(-\frac{\partial f}{\partial E_{\boldsymbol{k}}} \right) \right]^2}, \quad (12.3)$$

where $E_{\boldsymbol{k}}$ represents the energy dispersion of a hole-carrier, and $f(E_{\boldsymbol{k}}, \mu)$ the Fermi distribution function at energy $E_{\boldsymbol{k}}$ and chemical potential μ. By using the effective one-electron type energy band for the hole-carriers derived by Ushio and Kamimura for LSCO which was presented in Chap. 11 [112], we have calculated both the hole-concentration- and temperature- dependences of R_H. The calculated results of R_H in $La_{2-x}Sr_xCuO_4$ for $T = 80$ K and 300 K are shown as a function of x in Fig. 12.3, where the experimental results of R_H by Takagi et al. [10] are also shown for comparison. It is seen from this figure that the calculated results of R_H decrease like $1/x$ in very low concentrations. Then in underdoped to overdoped region it decreases more rapidly than the $1/x$ behaviour and at around $x \sim 0.3$ the Hall coefficient changes its sign from positive (hole-like) to a negative one (electron-like). This behaviour in the x dependence of R_H coincides well with the experimental results by Takagi et al. [10]. According to the present theoretical result, the reason for the sign change of R_H is due to the fact that, in the region of $0.3 \geq x$, the four small pockets of Fermi surface change into a large Fermi surface as shown in Figs. 12.4 and 12.5 and that, on the large Fermi surface the second derivatives of the energy dispersion such as $\frac{\partial^2 E_{\boldsymbol{k}}}{\partial k_y^2}$ change their sign over a dominant region of the Fermi surface. This leads to the negative Hall coefficient R_H at $T = 0$ K in the well overdoped concentration region.

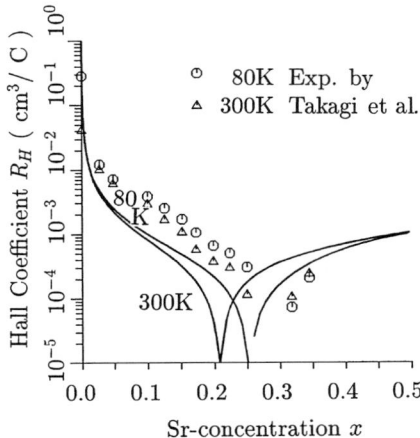

Fig. 12.3. The calculated concentration dependence of the Hall coefficient R_H for $T = 80\,\text{K}$ and $T = 300\,\text{K}$ based on the K–S model, together with the experimental results by Takagi et al. [10]

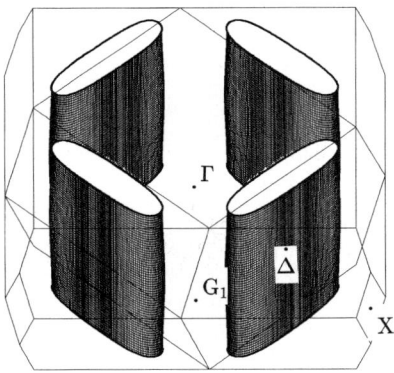

Fig. 12.4. The Fermi surface for $x = 0.15$ based on K–S model. Here two kinds of Brillouin zones are also shown. One at the outermost part is the ordinary Brillouin zone and the inner part is the folded Brillouin zone for the antiferromagnetic unit cell in LSCO. Here the k_x axis is taken along $\overline{\varGamma G_1}$, corresponding to the x-axis (the Cu–O–Cu direction) in a real space, where $\varGamma = (0,0,0)$ and $G_1 = (\pi/a, 0, 0)$

When temperature increases, the holes with higher energy than Fermi energy contribute more to the Hall coefficient of a negative sign. In other words, at temperatures higher than 300 K, the dominant number of holes lie on the states for which the second derivatives of the energy dispersion is negative. As a result the Hall coefficient becomes negative at higher temperatures. In this way the K–S model can explain the anomalous behaviour of the observed Hall effect without introducing any disposal parameter.

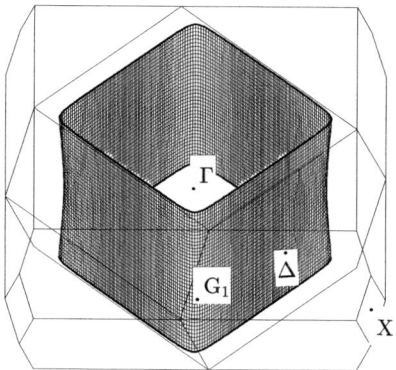

Fig. 12.5. The Fermi surface for $x = 0.35$ based on K–S model. Here two kinds of Brillouin zones are also shown. One at the *outermost* part is the ordinary Brillouin zone and the *inner* part is the folded Brillouin zone for the antiferromagnetic unit cell in LSCO. Here the k_x axis is taken along $\overline{\Gamma G_1}$, corresponding to the x-axis (the Cu-O-Cu direction) in a real space

As regards the temperature dependence of the Hall effect, we can also calculate the Hall angle θ_H as a function of temperature, where the Hall angle θ_H is defined by $\tan^{-1}(\sigma_{xy}/\sigma_{xx})$. Since $\cot\theta_H$ is given by

$$\cot\theta_H = \frac{1}{\sigma_{xx} R_H H} , \qquad (12.4)$$

where H is an external magnetic field, we have calculated the temperature-dependence of $\cot\theta_H$ by using the calculated results of (12.1), (12.2) and (12.3) for LSCO for several values of x in the underdoped region. We find that $\cot\theta_H$ is proportional to T^2. This is consistent with recent experiments of LSCO by Ando and his coworkers [181].

They have ascribed the peculiar T-dependence of $\cot\theta_H$ to the flat band near G_1 point. The flat band, however, is apart from Fermi energy in the underdoped region of LSCO and it is not probable that the peculiar T-dependence of $\cot\theta_H$ is due to the flat band near G_1 point. As will be described in the following section based on K–S model, however, the heavy mass band in the LF-phase contributes to the peculiar T-dependence of $\cot\theta_H$. Thus the peculiar T-dependence of $\cot\theta_H$ observed by Ando and his coworkers also supports the K–S model.

12.4 Electronic Entropy

Electronic entropy also exhibits an anomalous behaviour in its unusual doping dependence in $La_{2-x}Sr_xCuO_4$ (abbreviated as LSCO), as observed by Loram et al. [182]. According to their experimental results, when the hole concentration (x) increases, the electronic entropy increases and takes a maximum around $x = 0.25$. Then a question arises how one can explain the appearance of the maximum. Since the electronic entropy depends crucially on the density of states, it contains important information on the metallic state of cuprates. When we adopt the mean-field approximation for the K–S Hamiltonian, we can calculate the electronic entropy as functions of hole-concentration (x) and temperature (T) from the following well-known formula

$$S(T,x) = -k_\mathrm{B} \int_{-\infty}^{\infty} \Big[f(\varepsilon,\mu) \ln f(\varepsilon,\mu) \\ + \{1 - f(\varepsilon,\mu)\} \ln\{1 - f(\varepsilon,\mu)\} \Big] \rho(\varepsilon)\, d\varepsilon\,, \qquad (12.5)$$

where $\rho(\varepsilon)$ is the density of states function and $f(\varepsilon,\mu)$ the Fermi distribution function at energy ε and chemical potential μ.

In this context, Kamimura, Hamada and Ushio [183] calculated the x and T dependences of the electronic entropy $S(T,x)$ of LSCO for $T = 100\,\mathrm{K}$ and $200\,\mathrm{K}$, by inserting $\rho_\mathrm{KS}(\varepsilon)$ into $\rho(\varepsilon)$ in (12.5), where $\rho_\mathrm{KS}(\varepsilon)$ represents the density of states for the highest occupied band in LSCO calculated by Ushio and Kamimura [112], which was given as the #1 band in Fig. 11.1 in Chap. 11. The calculated electronic entropy functions of LSCO at $T = 100\,\mathrm{K}$ and $200\,\mathrm{K}$, $S_\mathrm{KS}(100,x)$ and $S_\mathrm{KS}(200,x)$, are shown by solid curves in Fig. 12.6, along with the measured values by Loram et al. [182] shown by filled squares for comparison. Reflecting the appearance of a saddle point singularity in

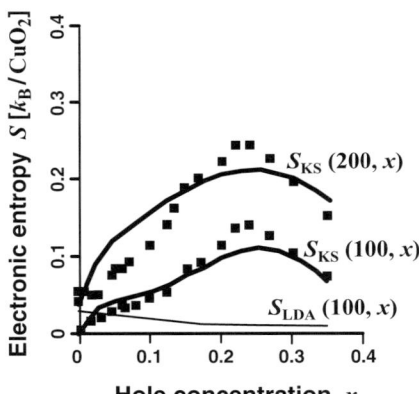

Fig. 12.6. The calculated electronic entropies of LSCO at $T = 100\,\mathrm{K}$ and $T = 200\,\mathrm{K}$, $S_\mathrm{KS}(100,x)$, $S_\mathrm{KS}(200,x)$, based on the K–S model (*solid curves*), and that of the "LDA state" at $T = 100\,\mathrm{K}$, $S_\mathrm{LDA}(100,x)$, based on LDA density of states (*thin solid line*). The experimental results of Loram et al. [182] are also shown by closed squares for comparison. Note that both the calculated values and the experimental data for the 200 K entropy have been shifted to aid clarity

12.4 Electronic Entropy

$\rho_{\mathrm{KS}}(\varepsilon)$ at $x = 0.3$ shown in Fig. 11.10, the electronic entropy increases up to around $x = 0.25$, where both the calculated and measured entropy reach a broad maximum. The agreement between the calculated entropy functions at 100 K and 200 K, $S_{\mathrm{KS}}(100, x)$ and $S_{\mathrm{KS}}(200, x)$, and the experimental results by Loram et al. [182] are a strong vindication of the theoretical treatment of the K–S model. Here it should be emphasized that there are no adjustable parameters in this calculation, and the doping dependence of the chemical potential μ, i.e., $\mu(x)$, has also been explicitly taken into account. For comparison we have also calculated the entropy function using the density of states function, $\rho_{\mathrm{LDA}}(\varepsilon)$, obtained from the conduction band calculated by Shima et al. [106] in the local density approximation (LDA). By repeating the calculation, inserting $\rho_{\mathrm{LDA}}(\varepsilon)$ for $\rho(\varepsilon)$ in (12.5), we have generated the LDA entropy function, $S_{\mathrm{LDA}}(100, x)$. The calculated $S_{\mathrm{LDA}}(100, x)$ is shown as a thin solid line in Fig. 12.6 as a function of hole concentration x at $T = 100$ K. It is clearly seen that the calculated entropy function based on the LDA band is too small, and has no maximum, exposing the inadequacy of the electronic band structure based on the LDA band, even in the overdoped region. The lack of agreement between experimental results and the LDA entropy function on the one hand, and the rather good agreement between the experimental results and the K–S entropy function on the other, confirms the importance of including the electron correlation and local lattice distortion of CuO_6 octahedrons in any computational models for the cuprates. These many-body and lattice distortion coupled effects have been taken into account in the K–S model but not in LDA. Here we call the latter LDA band the "LDA state". The present results also suggest a significant mass enhancement of the hole-carriers. We can estimate that the effective mass of itinerant holes is about six times the free electron mass. Since this mass enhancement of hole-carriers is partly due to the effects of local lattice distortion based on the anti-Jahn–Teller distortion, we expect a large isotope effect in superconductivity, such as that observed in $HgBa_2CuO_4O_8$ and $La_{1.94}Sr_{0.06}CuO_4$ reported by Mülller [184], which we will describe in Chap. 14. The agreement between the K–S entropy function and experiments leads us to conclude that the metallic state described by the K–S model is appropriate for the cuprates in the normal state.

As we have described a number of times so far, the metallic regions on the K–S model are inhomogeneous, because the spin-correlated regions which trigger the metallic regions are distributed randomly over the CuO_2 planes. Further, the metallic regions are widely expanded by the dynamical effects of the 2D spin-fluctuation in the 2D AF Heisenberg spin system, compared with the length of the spin-correlated regions, as we described in detail in Sect. 11.2. Further we may say that the hole-carriers in the metallic state in the overdoped region have partly the nature of a large polaron with the six times heavier effective mass of a free electron mass.

12.5 Validity of the K–S Model in the Overdoped Region and Magnetic Properties

When the hole-carrier concentration increases in the overdoped region beyond the optimum doping concentration, the occupation rate of hole-carriers at the oxygen sites in a CuO_2 plane also increases. As a result, the number of O^{2-} closed shell configurations in a CuO_2 plane decreases. This means that superexchange interaction between localized spins via intervening O^{2-} ions in a CuO_2 plane is partly destroyed, and the effective superexchange interactions between the localized spins at Cu sites become weak. In this circumstance the local AF order in the underdoped region diminishes beyond the optimum doping and disappears at a certain hole concentration x_c in the overdoped region. In this context we may say that a kind of Néel temperature related to the local AF order in a spin-correlated region is dependent on the hole-concentration x, and thus we denote it by $\hat{T}_N(x)$. Since $\hat{T}_N(x)$ decreases with increasing x in the overdoped region beyond the optimum doping x_{opt}, it vanishes at x_c. Thus the K–S model does not hold in the hole concentration beyond x_c. In order to investigate the magnetic behaviour in the normal state, detailed experimental investigations for the temperature- and hole-concentration dependences of the spin susceptibility have been carried out by various experimental groups [10, 185, 186, 187, 188, 189] for the normal state of LSCO. According to these experimental results, the observed spin susceptibility χ_m shows the anomalous temperature dependence in the underdoped region in such a way that χ_m exhibits a broad maximum as a certain temperature T_{max}, indicating a two-dimensional (2D) antiferromagnetic (AF) behaviour due to the localized spins around Cu sites below T_{max}. Miyashita [190], and Okabe and Kikuchi [191] obtained such anomalous temperature dependence theoretically by performing quantum Monte Carlo simulation for the $S = 1/2$ 2D AF Heisenberg spin system on the square lattice of a finite size. Experimentally, however, this maximum disappears in the overdoped region beyond $x \approx 0.2 (= x_c)$, suggesting the suppression of the in-plane Cu–Cu superexchange interaction in a CuO_2 plane. This behaviour is consistent with the prediction by the K–S model that the superexchange interaction via the intervening O^{2-} ions becomes ineffective at x_c when the hole concentration increases in the overdoped region. As a result the shape of the Fermi surfaces changes from small ones to a larger one with a heavy effective mass, and the K–S model does not hold in the overdoped region beyond x_c. Thus the spin susceptibility changes from 2D antiferromagnetic behaviour to Pauli-like behaviour. Here we should remark that the existence of a flat band of the heavy effective mass with a large FS may be an origin of the T^2 dependence of $\cot\theta_H$ observed by Ando et al. [181], as we mentioned in Sect. 12.3.

A similar situation also appears even in the underdoped region when the temperature increases for a fixed hole concentration in the underdoped region. In other words, when the temperature increases for a certain hole concentration x_0 in the underdoped region in which the K–S model holds, the Fermi surfaces change from small ones to a larger one when temperature T exceeds $\hat{T}_N(x_0)$, at which the local AF order disappears. As a result, the spin susceptibility is expected to change from a 2D AF behaviour to a Pauli-like behaviour. Based on this argument, we can define the high energy pseudogap from the standpoint of the K–S model. In the following section we will discuss an origin of the high-energy pseudogap based on the K–S model not only from a qualitative standpoint but also from a quantitative standpoint.

12.6 The Origin of the High-Energy Pseudogap

12.6.1 Introduction

In connection with the anomalous hole-concentration dependences of the electronic entropy and of static spin susceptibility described in Sects. 12.3 and 12.4, Loram and his coworkers [182] proposed the concept of the "pseudogap" from the standpoint of the Fermi liquid picture, by attributing the anomalous behaviours to the reduction of the density of states near the Fermi energy, called the pseudogap. On the other hand, Nakano and his coworkers [192] found from their measurements of magnetic susceptibility and electrical resistivity for LSCO that there are two crossover lines, $T_{\max}(x)$ and $T^*(x)$ ($T_c < T^* < T_{\max}$), in the T-x phase diagram of cuprates with a superconducting transition temperature T_c, both of which decrease monotonically with increasing hole concentration x, as shown in Fig. 12.7. The upper crossover line $T_{\max}$ represents the temperature below which the magnetic susceptibility exhibits a broad peak, arising from the gradual development of 2D antiferromagnetic spin correlation, while the lower crossover line T^* represents the temperature below which a spin gap may open up in the magnetic excitation spectrum around $q = (\pi, \pi)$. Now $T_{\max}$ and T^* are called the "high-energy" and "low-energy" pseudogaps, respectively.

In this section we will discuss the origin of the high-energy pseudogap on the basis of the K–S model. In the course of calculating the high-energy pseudogap, we will discuss again the electronic entropy. Although we have shown in Sect. 12.4 that the anomalous concentration dependence of the electronic entropy can be explained successfully by the K–S model over the wide range of hole-concentration from underdoped to well-overdoped region without introducing any adjustable parameters, the argument in Sect. 12.4 has clarified that the K–S model does not hold in the concentration beyond x_c, where the shape of the Fermi surface changes from small ones to a large one due to the disappearance of local AF order. Thus we have to reinvestigate the behaviour of the electronic entropy in the overdoped region beyond x_c. We will show on the basis of the K–S model that the origins of both $T_{\max}$ and the strange hole-concentration-dependence of the electronic entropy are explained by a unified mechanism of phase change between the phase consisting of small

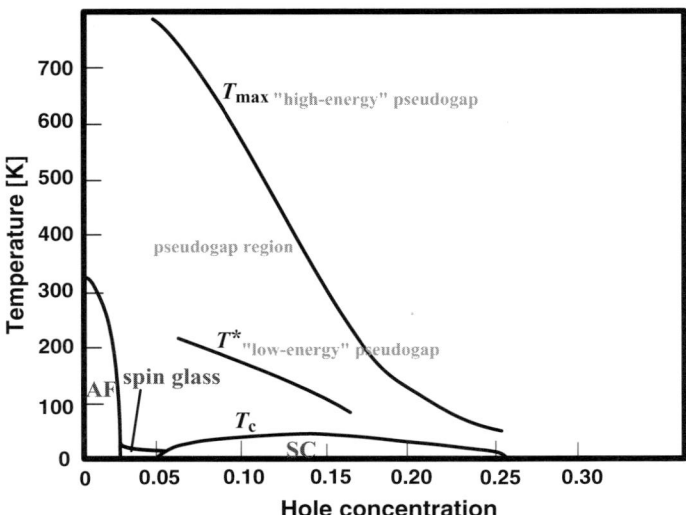

Fig. 12.7. The phase diagram of temperature T and hole concentration x for LSCO, showing the "high-energy" pseudogap $T_{\max}$ and the "low-energy" pseudogap T^*. Superconducting transition temperature T_c is also shown (After [49])

Fermi surfaces (abbreviated as "SF-phase") and the phase consisting of a large Fermi surface (abbreviated as "LF-phase").

12.6.2 Calculation of Free Energies of the SF- and LF-Phases

According to the description mentioned in Sect. 12.4, the electronic structure of LSCO changes from the SF-phase to the LF-phase at x_c in the overdoped region when x increases. Similarly, when the temperature increases for a fixed hole concentration in the underdoped regime, the local AF order is destroyed so that the SF-phase is considered to change to the LF-phase at a certain temperature T_0. We consider that T_0 may correspond to $T_{\max}$ in the experiments by Nakano et al. [192].

When we write the difference between the free energies per Cu ion of the LF- and SF-phases as

$$\Delta F(T,x) = F_{\mathrm{LF}}(T,x) - F_{\mathrm{SF}}(T,x) , \qquad (12.6)$$

$T_{\max}$ and the critical concentration x_c are defined as $\Delta F(T_{\max}, x) = 0$ and $\Delta F(T=0, x_c) = 0$, respectively. First the internal energy in the SF-phase is expressed in the following way;

$$E_{\mathrm{SF}}(T,x) = E_{\mathrm{SF}}(0,0) + E_{\mathrm{kin}}^{(\mathrm{SF})}(T,x) - E_{\mathrm{kin}}^{(\mathrm{SF})}(0,0) , \qquad (12.7)$$

where $E_{\mathrm{SF}}(0,0)$ and $E_{\mathrm{kin}}^{(\mathrm{SF})}(T,x)$ represent, respectively, the internal energy per Cu ion in a system of the local antiferromagnetic ordering with $T = 0\,\mathrm{K}$

12.6 The Origin of the High-Energy Pseudogap

and $x = 0$ and the kinetic energy per Cu ion in a hole-carrier system with concentration x at T. Since $E_{\rm SF}(0,0)$ includes the kinetic energy, we have to subtract $E_{\rm kin}^{({\rm SF})}(0,0)$ from $E_{\rm kin}^{({\rm SF})}(T,x)$ in order to avoid the double counting. By using the density of states per Cu ion for the highest energy band calculated by Ushio and Kamimura [112], $\rho_{\rm KS}(\varepsilon)$, and the Fermi distribution function $f(\varepsilon, \mu(x))$, $E_{\rm kin}^{({\rm SF})}(T,x)$ can be expressed as

$$E_{\rm kin}^{({\rm SF})}(T,x) = \int_{-\infty}^{\infty} \varepsilon\, \rho_{\rm KS}(\varepsilon)\, f(\varepsilon, \mu(x))\, d\varepsilon\,, \tag{12.8}$$

where we consider explicitly the x-dependence of the chemical potential μ as $\mu(x)$. The entropy per Cu ion in the SF-phase is calculated by a well-known formula

$$S_{\rm SF}(T,x) = -k_B \int_{-\infty}^{\infty} \Big[f(\varepsilon, \mu(x)) \ln f(\varepsilon, \mu(x)) \\ + \{1 - f(\varepsilon, \mu(x))\} \ln\{1 - f(\varepsilon, \mu(x))\} \Big] \rho_{\rm KS}(\varepsilon)\, d\varepsilon\,. \tag{12.9}$$

Thus the free energy per Cu ion is calculated by the formula,

$$F_{\rm SF}(T,x) = E_{\rm SF}(T,x) - T S_{\rm SF}(T,x)\,. \tag{12.10}$$

Similarly we denote the internal energy per Cu ion in the LF-phase by $E_{\rm LF}(T,x)$. In the LF-phase the local antiferromagnetic ordering does not exist so that we first assume that the electronic system may be treated by an ordinary band theory by the local density approximation (LDA) with the $(1+x)$ hole concentration. Thus $E_{\rm LF}(T,x)$ can be expressed as

$$E_{\rm LF}(T,x) = E_{\rm LF}(0,0) + E_{\rm kin}^{({\rm LF})}(T,x) - E_{\rm kin}^{({\rm LF})}(0,0)\,, \tag{12.11}$$

where a conduction band corresponds to a b_{1g}^* energy band with the main character of Cu $d_{x^2-y^2}$ orbital in the LDA band [147]. The $E_{\rm LF}(0,0)$ represents the internal energy per Cu ion in the LF-phase with $T = 0\,{\rm K}$ and $x = 0$, in which the b_{1g}^* energy band is half-filled. The entropy per Cu ion in the LF-phase is calculated by

$$S_{\rm LF}(T,x) = -k_B \int_{-\infty}^{\infty} \Big[f(\varepsilon, \mu(x)) \ln f(\varepsilon, \mu(x)) \\ + \{1 - f(\varepsilon, \mu(x))\} \ln\{1 - f(\varepsilon, \mu(x))\} \Big] \rho_{\rm LDA}(\varepsilon)\, d\varepsilon\,, \tag{12.12}$$

where $\rho_{\rm LDA}(\varepsilon)$ is the density of states for a system with $(1+x)$ hole concentration of the b_{1g}^* energy band. Then the free energy per Cu ion in the LF-phase is given by

$$F_{\rm LF}(T,x) = E_{\rm LF}(T,x) - T S_{\rm LF}(T,x)\,. \tag{12.13}$$

The calculated results of $S_{\rm SF}(T=100\,{\rm K},x)$ and $S_{\rm SF}(T=200\,{\rm K},x)$, which have been calculated using $\rho_{\rm KS}(\varepsilon)$ for the K–S model, have been already shown in Fig. 12.6. In this figure we have also shown $S_{\rm LF}(100,x)$ as $S_{\rm LDA}(100,x)$ for "LDA state". Since we have seen in Fig. 12.6 that $S_{\rm LF}(100,x)$ is very small compared with experimental results shown by the solid squares, we have concluded that the electronic state in the overdoped region beyond $x_{\rm c}=0.25$ can not be expressed by the "LDA state" represented by the ordinary LDA bands. In this context, we have modified the density of states of the LDA b_{1g}^* band so as to reproduce the experimental values of electronic entropy in the overdoped region above $x_{\rm c}$. The density of states calculated by Ushio and Kamimura in the SF-phase, $\rho_{\rm KS}(\varepsilon)$, and the modified density of states in the LF-phase which we denote $\rho_{\rm LDA}^*(\varepsilon)$ are shown in Figs. 12.8(a) and (b), respectively. The modified density of states at a van Hove singularity which lies at the center of the band is about 6 times larger than that of the LDA band. We call the new state the "modified LDA state". Since the observed maxima in the x-dependence of entropy in LSCO appear around $x=0.25$ both for $T=100\,{\rm K}$ and $200\,{\rm K}$, we adopt $x_{\rm c}=0.25$ [14]. By using the modified density of states $\rho_{\rm LDA}^*(\varepsilon)$ for $\rho_{\rm LDA}(\varepsilon)$ in (12.12), we have recalculated the electronic entropies for $T=100\,{\rm K}$ and $200\,{\rm K}$ for the concentration region beyond $x_{\rm c}$, which are denoted by $S_{\rm LDA}^*(100,x)$ and $S_{\rm LDA}^*(200,x)$, respectively. The calculated results of $S_{\rm LDA}^*(100,x)$ and $S_{\rm LDA}^*(200,x)$ for $x>x_{\rm c}$ are shown in Fig. 12.9 together with $S_{\rm KS}(100,x)$ and $S_{\rm KS}(200,x)$. Experimental results by Loram et al. are also shown by closed squares in Fig. 12.9. As seen in this figure, we can explain successfully the observed x-dependence of the electronic entropies both in the underdoped and overdoped region, by defining the effective mass of a dopant hole in the overdoped region to be six times heavier than the free electron mass. Although the discontinuity appears around $x_{\rm c}$ in Fig. 12.9, this should be

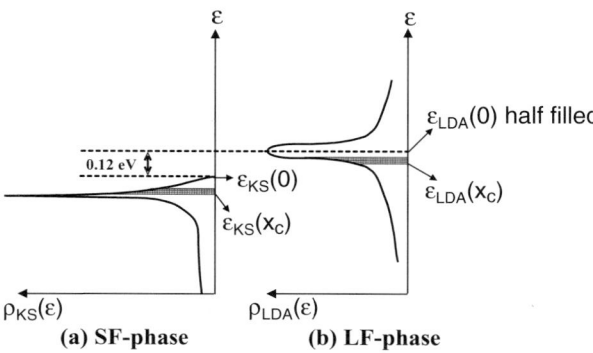

Fig. 12.8. (a) The density of states calculated by Ushio-Kamimura $\rho_{\rm KS}(\varepsilon)$ in the SF-phase. (b) The modified density of states for the b_{1g}^* band $\rho_{\rm LDA}^*(\varepsilon)$ in the LF-phase, so as to reproduce the observed electronic entropy by Loram et al. [193]

12.6 The Origin of the High-Energy Pseudogap

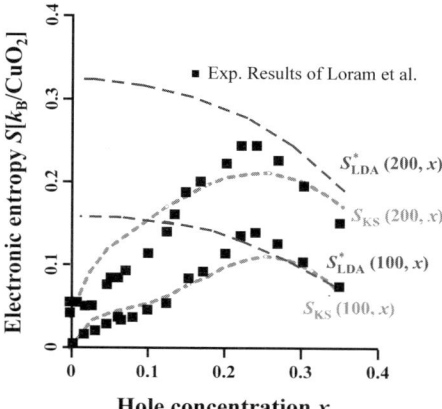

Fig. 12.9. The calculated electronic entropies of LSCO at $T = 100\,\text{K}$ and $T = 200\,\text{K}$, $S_{\text{KS}}(100, x)$, $S_{\text{KS}}(200, x)$, based on the K–S model, and that of the "modified LDA state" at $T = 100\,\text{K}$ and $200\,\text{K}$, $S^*_{\text{LDA}}(100, x)$ and $S^*_{\text{LDA}}(200, x)$, based on LDA density of states. The experimental results of Loram et al. [182] are also shown by closed squares for comparison. Note that both the calculated values and the experimental data for the 200 K entropy have been shifted to aid clarity

smeared out by taking account of the broadening effect of Fermi surfaces in the SF-phase. From this result we conclude that the appearance of the maximum in the observed entropy may be considered as an experimental evidence for a phase change from the SF-phase to the LF-phase at x_c. Further, from the modified density of states $\rho^*_{\text{LDA}}(\varepsilon)$ for the "modified LDA state", we conclude that the effective mass of hole carriers are about 6 times heavier than the free electron mass in the overdoped region. Since the effective mass of the hole carriers in the highest occupied (#1) band in the underdoped region in Fig. 11.1 is about ten times heavier than the free electron mass [112], we may say that the hole carriers in the superconducting cuprates behave like large polarons with heavy mass.

12.6.3 Origin of the "High-Energy" Pseudogap

By using the density of states of the highest occupied (#1) band [112], $\rho_{\text{KS}}(\varepsilon)$, for the underdoped to overdoped region below x_c and the modified density of states for the b^*_{1g} band in the "modified LDA state", $\rho^*_{\text{LDA}}(\varepsilon)$, for the overdoped region above x_c, $T_{\max}$ is calculated from the following equations;

$$\Delta F(T_{\max}, x) = 0 , \qquad (12.14)$$

where

$$\begin{aligned}
\Delta F(T,x) &= F^*_{\mathrm{LF}}(T,x) - F_{\mathrm{SF}}(T,x) \\
&= \{E^*_{\mathrm{LF}}(T,x) - TS^*_{\mathrm{LF}}(T,x)\} - \{E_{\mathrm{SF}}(T,x) - TS_{\mathrm{SF}}(T,x)\} \\
&= E^*_{\mathrm{LF}}(0,0) - E_{\mathrm{SF}}(0,0) + \{E^{*(\mathrm{LF})}_{\mathrm{kin}}(T,x) - E^{*(\mathrm{LF})}_{\mathrm{kin}}(0,0)\} \\
&\quad - \{E^{(\mathrm{SF})}_{\mathrm{kin}}(T,x) - E^{(\mathrm{SF})}_{\mathrm{kin}}(0,0)\} - T\{S^*_{\mathrm{LF}}(T,x) - S_{\mathrm{SF}}(T,x)\} \\
&= E_{\mathrm{AF}} + \int_{-\infty}^{\infty} \varepsilon\, \rho^*_{\mathrm{LDA}}(\varepsilon)\, f(\varepsilon,\mu(x))\, d\varepsilon - \int_{-\infty}^{\varepsilon_{\mathrm{LDA}}(0)} \varepsilon\, \rho^*_{\mathrm{LDA}}(\varepsilon)\, d\varepsilon \\
&\quad - \left\{ \int_{-\infty}^{\infty} \varepsilon\, \rho_{\mathrm{KS}}(\varepsilon)\, f(\varepsilon,\mu(x))\, d\varepsilon - \int_{-\infty}^{\varepsilon_{\mathrm{KS}}(0)} \varepsilon\, \rho_{\mathrm{KS}}(\varepsilon)\, d\varepsilon \right\} \\
&\quad - T\{S^*_{\mathrm{LF}}(T,x) - S_{\mathrm{SF}}(T,x)\}\,. \quad (12.15)
\end{aligned}$$

Here $E_{\mathrm{AF}} \equiv E^*_{\mathrm{LF}}(0,0) - E_{\mathrm{SF}}(0,0)$. Further $\varepsilon_{\mathrm{KS}}(0)$ and $\varepsilon_{\mathrm{LDA}}(0)$ represent the band energies of the highest occupied (#1) [112] band and of the "modified b^*_{1g} band" for $x=0$, respectively, as shown in Figs. 12.8(a) and (b). $T_{\max}$ is calculated from (12.14) and (12.15) as a function of x. In doing so, the value of E_{AF} is found to be 0.043 eV by inserting $T_{\max} = 1000\,\mathrm{K}$ at $x=0$ into (12.14) and (12.15), where $T_{\max} = 1000\,\mathrm{K}$ at $x=0$ is obtained by extrapolating the experimental data by Nakano and his coworkers [188, 192] to $x=0$. The E_{AF} thus determined corresponds to the energy change from the half-filled band in the LF-phase to a Mott–Hubbard insulator in the SF-phase. Furthermore, information about the energy difference, $\varepsilon_{\mathrm{LDA}}(0) - \varepsilon_{\mathrm{KS}}(0)$, is necessary. We determine it by requiring that $\Delta F(T=0\,\mathrm{K},x)$ vanishes at x_c. As for x_c, 0.25 is chosen from the concentration at which the entropy shows a maximum in the experimental data by Loram et al. [182]. Then we have determined $\varepsilon_{\mathrm{LDA}}(0) - \varepsilon_{\mathrm{KS}}(0)$ to be 0.12 eV. Using $E_{\mathrm{AF}} = 0.043$ eV and $\varepsilon_{\mathrm{LDA}}(0) - \varepsilon_{\mathrm{KS}}(0) = 0.12$ eV, we have calculated the x-dependence of $T_{\max}$ from (12.14) and (12.15) quantitatively. The calculated result is shown in Fig. 12.10. When the calculated x-dependence of $T_{\max}$ is compared with that observed for LSCO by Nakano et al. [192], shown in the inset of Fig. 12.10, we find that the agreement is fairly good. Thus, from the standpoint of the K–S model one may say that the phase change between the SF- and LF-phases corresponds to the "high-energy" pseudogap, $T_{\max}$. As shown in the phase diagram of Fig. 12.7, we may call a region below the curve of $T_{\max}$ the "pseudogap region", where SF- and LF-phases coexist. Because of this coexistence of two phases, we expect that the temperature dependences of spin susceptibility and Fermi surface are conspicuous.

In this section we have shown from the calculations of the electronic entropies in the SF- and LF-phases and also from the calculated free energy difference between the SF- and LF-phases based on the K–S model that both the observed peculiar x-dependences of entropy by Loram et al. and of the observed $T_{\max}$ by Nakano et al. can be explained by a unified mechanism of a phase change between the SF- and LF-phases in the K–S model. Further, by fitting the calculated x-dependence of the electronic entropy to the one

12.6 The Origin of the High-Energy Pseudogap

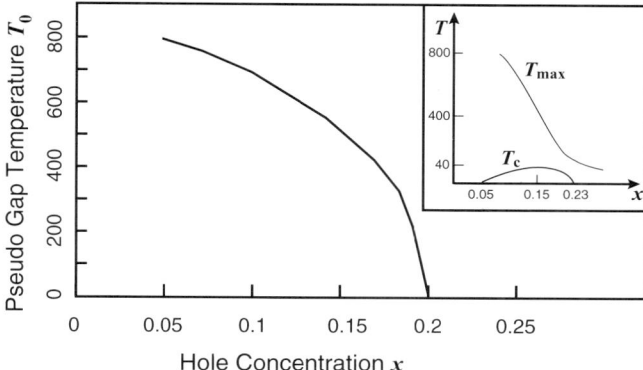

Fig. 12.10. The calculated results for the hole concentration dependence of T_{max} in LSCO. The experimental data for T_{max} by Nakano et al. [192] are also shown, together with the sketch of the x dependence of T_c in the inset

observed in the overdoped region beyond $x_c = 0.25$, we may understand that the hole carriers in cuprates behave like large polarons with an effective mass of about six times heavier than the free electron mass. We can say that the origin of the heavy effective mass is due to the interplay between the electron correlation and the local lattice distortion of CuO_6 octahedrons. Finally we would like to make brief comments on the origin of "low-energy pseudogap" T^* shown in Fig. 12.7. According to the K–S model, T^* corresponds to the spin-excitation-energy in the metallic region of a finite size. We can estimate this energy from the spin-wave excitation spectrum around the edge of the AF Brillouin zone calculated by Suzuki and Kamimura for the magnon spectra in K_2NiF_4 [194].

13 Electron–Phonon Interaction and Electron–Phonon Spectral Functions

13.1 Introduction

In 1996, by choosing LSCO as an object of study, Kamimura, Matsuno, Suwa and Ushio [30] showed quantitatively on the basis of the K–S model that the electron–phonon interactions in cuprates are very strong and that the interplay between the electron–phonon interaction and local antiferromagnetic (AF) order gives rise to the Cooper pairs of $d_{x^2-y^2}$ symmetry. This result was the first theoretical support for the original idea by Bednorz and Müller in their discovery of high temperature superconductivity, that strong electron–phonon interactions in oxides may lead to the invention of new superconducting materials which can go beyond the superconductors governed by the BCS theory. In 2001 Lanzara and her coworkers [195] presented experimental evidence for strong electron–phonon coupling in cuprates by investigating the electronic quasiparticle dispersions in three different families of Bi2212 with use of the angle-resolved photoemission spectroscopy (ARPES). In this context, we will describe in detail in this chapter the method of how to calculate the electron–phonon interactions and the electron–phonon spectral functions in the K–S model following the theory of high temperature superconductivity developed by Kamimura and his coworkers [26, 30].

13.2 Calculation of the Electron–Phonon Coupling Constants for the Phonon Modes in LSCO

All relevant properties of the electron–phonon systems, including superconductivity, are derived from the electron–phonon spectral functions $\alpha^2 F$, which are defined as follows,

$$\alpha^2 F_{\uparrow\uparrow}(\Omega, \mathbf{k}, \mathbf{k}') = \rho(E_\mathrm{F}) \sum_\gamma \frac{V_\uparrow^\gamma(\mathbf{k}, \mathbf{k}') V_\uparrow^\gamma(-\mathbf{k}, -\mathbf{k}')}{2\omega_{\mathbf{q}}^\gamma} \delta(\Omega - \omega_{\mathbf{k}'-\mathbf{k}}^\gamma) \,, \quad (13.1)$$

$$\alpha^2 F_{\uparrow\downarrow}(\Omega, \mathbf{k}, \mathbf{k}') = \rho(E_\mathrm{F}) \sum_\gamma \frac{V_\uparrow^\gamma(\mathbf{k}, \mathbf{k}') V_\downarrow^\gamma(-\mathbf{k}, -\mathbf{k}')}{2\omega_{\mathbf{q}}^\gamma} \delta(\Omega - \omega_{\mathbf{k}'-\mathbf{k}}^\gamma) \,. \quad (13.2)$$

Here $\alpha^2 F_{\uparrow\uparrow}(\Omega, \mathbf{k}, \mathbf{k}')$ and $\alpha^2 F_{\uparrow\downarrow}(\Omega, \mathbf{k}, \mathbf{k}')$ are the spectral functions which are related to the processes of virtual emission and absorption of various modes of phonons by the interaction with a single electron and to the scattering processes of a pair of electrons from a pair state $(\mathbf{k}\uparrow, -\mathbf{k}\downarrow)$ to a different pair state $(\mathbf{k}'\uparrow, -\mathbf{k}'\downarrow)$, respectively. Further, $\rho(E_F)$ is the density of states at the Fermi energy E_F. The electron–phonon interaction matrix element between the $\mathbf{k}$ and $\mathbf{k}'$ states with spin σ, $V_\sigma^\gamma(\mathbf{k}, \mathbf{k}')$, is defined as follows:

$$H_{e-p} = \sum_{\mathbf{K}\mathbf{k}\mathbf{k}'\mathbf{q}\gamma\sigma} \frac{V_\sigma^\gamma(\mathbf{k}, \mathbf{k}')}{\sqrt{N\omega_\mathbf{q}^\gamma}} c_{\mathbf{k}\sigma}^\dagger c_{\mathbf{k}'\sigma} \left(b_{\mathbf{q}\gamma} + b_{-\mathbf{q}\gamma}^\dagger\right) \delta_{\mathbf{k},\mathbf{k}'+\mathbf{q}+\mathbf{K}}, \quad (13.3)$$

where $b_{\mathbf{q}\gamma}$ is an annihilation operator of phonon mode γ with momentum $\mathbf{q}$, $\omega_\mathbf{q}^\gamma$ the phonon frequency of the wave vector $\mathbf{q}$ in the AF Brillouin zone, N the total number of AF unit cells in a crystal, and $\delta_{\mathbf{k},\mathbf{k}'+\mathbf{q}+\mathbf{K}}$ takes the value 1 only when $\mathbf{k} - \mathbf{k}' - \mathbf{q}$ coincides with a reciprocal lattice vector in the AF Brillouin zone, $\mathbf{K}$, and 0 for other cases. The spectral function $\alpha^2 F_{\uparrow\uparrow}(\Omega, \mathbf{k}, \mathbf{k}')$ causes a mass enhancement of an electron near the Fermi surface due to the electron–phonon interaction and a finite lifetime of quasi-particle states. On the other hand, the spectral function $\alpha^2 F_{\uparrow\downarrow}(\Omega, \mathbf{k}, \mathbf{k}')$ contributes to the formation of the Cooper-pair with spin-singlet. These two kinds of spectral functions in (13.1) and (13.2) are different from each other in the present case due to the fact that the wave function for up-spin carriers differs from that for down-spin carriers, although they are the same in the ordinary BCS case. Frequently this electron–phonon spectral function is averaged over either one of $\mathbf{k}$ and $\mathbf{k}'$ or over both of the $\mathbf{k}$ and $\mathbf{k}'$ values in the electron states $(\mathbf{k}, \mathbf{k}')$ on the Fermi surface, as is shown below,

$$\alpha^2 F_{\uparrow\downarrow}(\Omega, \mathbf{k}) = \frac{1}{\rho(E_F)} \sum_{\mathbf{k}'} \alpha^2 F_{\uparrow\downarrow}(\Omega, \mathbf{k}, \mathbf{k}') \delta(E_{\mathbf{k}'}^0 - E_F), \quad (13.4)$$

$$\alpha^2 F_{\uparrow\downarrow}(\Omega) = \frac{1}{\rho(E_F)^2} \sum_{\mathbf{k}\mathbf{k}'} \alpha^2 F_{\uparrow\downarrow}(\Omega, \mathbf{k}, \mathbf{k}') \delta(E_{\mathbf{k}}^0 - E_F) \delta(E_{\mathbf{k}'}^0 - E_F). \quad (13.5)$$

Here we pay attention to the spectral function for a spin–singlet defined on the Fermi surface and averaged over k_z-axis. Such a spectral function is denoted by $\alpha^2 F_{\uparrow\downarrow}(\Omega, \theta, \theta')$, which is defined as follows,

$$\alpha^2 F_{\uparrow\downarrow}(\Omega, \theta, \theta') = \frac{1}{\rho(E_F)^2 N} \sum_{\mathbf{k}\mathbf{k}'\mathbf{q}} \alpha^2 F_{\uparrow\downarrow}(\Omega, \mathbf{k}, \mathbf{k}')$$
$$\times \delta_{\mathbf{k},\mathbf{k}'+\mathbf{q}+\mathbf{K}} \delta(E_{\mathbf{k}}^0 - E_F) \delta(E_{\mathbf{k}'}^0 - E_F)$$
$$\times \delta(\theta - \tan^{-1}\frac{k_y}{k_x}) \delta(\theta' - \tan^{-1}\frac{k_y'}{k_x'}). \quad (13.6)$$

Here $\rho(E_F)$ and $E_{\mathbf{k}}^0$ are the density of states of hole carriers at the Fermi energy and the energy of the many-body-effect included band dispersion at a wave-vector $\mathbf{k}$, respectively, both of which have been calculated in Chap. 11.

13.2 Calculation of the Electron–Phonon Coupling Constants

Following the method of Motizuki, Suzuki and Shirai [156, 157, 158], we will derive the expression of a spectral function in the tight binding form in the next section.

In this section we describe the formalisms of how to calculate the spectral functions, based on the many-body-effect included tight binding Hamiltonian (11.1) given in Chap. 11. With the use of the electron–phonon coupling constant $V_\sigma^\gamma(\mathbf{k}, \mathbf{k}')$ defined in (13.3), the momentum-dependent spectral function for a singlet Cooper pair, $\alpha^2 F_{\uparrow\downarrow}(\Omega, \theta, \theta')$, is expressed as follows,

$$\alpha^2 F_{\uparrow\downarrow}(\Omega, \theta, \theta') = \frac{1}{\rho(E_F)} \sum_{\mathbf{k}\mathbf{k}'\mathbf{q}} \sum_\gamma \frac{V_\uparrow^\gamma(\mathbf{k},\mathbf{k}') V_\downarrow^\gamma(-\mathbf{k},-\mathbf{k}')}{2N\omega_\mathbf{q}^\gamma}$$

$$\times \delta_{\mathbf{k},\mathbf{k}'+\mathbf{q}+\mathbf{K}} \delta(E_\mathbf{k}^0 - E_F) \delta(E_{\mathbf{k}'}^0 - E_F) \delta(\Omega - \omega_{\mathbf{k}'-\mathbf{k}}^\gamma)$$

$$\times \delta\left(\theta - \tan^{-1}\frac{k_y}{k_x}\right) \delta\left(\theta' - \tan^{-1}\frac{k_y'}{k_x'}\right) . \quad (13.7)$$

Now we calculate the momentum-dependent spectral function by averaging it with respect to phonon frequency $\omega_\mathbf{q}^\gamma$; in other words by replacing $\delta(\Omega - \omega_{\mathbf{k}'-\mathbf{k}}^\gamma)$ by the phonon density of states $P(\Omega)$. The obtained expression is the following:

$$\alpha^2 F_{\uparrow\downarrow}(\Omega, \theta, \theta') = \rho(E_F) N \sum_\gamma \frac{\langle V_\uparrow^\gamma(\mathbf{k}\mathbf{k}') V_\downarrow^\gamma(-\mathbf{k}-\mathbf{k}') \rangle_{av.}}{2\Omega} P(\Omega) . \quad (13.8)$$

Here $\langle \cdots \rangle_{av.}$ means the average over k_z and k_z' on the Fermi surfaces, where $k_y/k_x = \tan\theta$ and $k_y'/k_x' = \tan\theta'$.

In order to obtain the electron–phonon interaction, we calculate the change of the energy bands when the ions are displaced by a small amount $\delta\mathbf{R}_{l\mu}$ from their equilibrium positions $\mathbf{R}_{l\mu}$. Following the method of Motizuki et al. [156, 157, 158], we adopt the Fröhlich approach, in which one assumes that the atomic wave functions are not changed when the ions are displaced by a small amount. Thus we use the atomic wave functions which move rigidly with ions, in calculating the energy bands for a displaced structure. Therefore, the basis function in the displaced structure becomes $\varphi_a(\mathbf{r} - \mathbf{R}_{l\mu} - \delta\mathbf{R}_{l\mu})$ and the Bloch function in the displaced structure is constructed as follows:

$$\Phi_{\mu a \mathbf{k}}(\mathbf{r}) = \frac{1}{\sqrt{N}} \sum_l e^{i\mathbf{k}\cdot\mathbf{R}_{l\mu}} \varphi_a(\mathbf{r} - \mathbf{R}_{l\mu} - \delta\mathbf{R}_{l\mu}) , \quad (13.9)$$

where $\mathbf{R}_{l\mu} = \mathbf{R}_l + \boldsymbol{\tau}_\mu$ represents the position of the μth ion in the lth unit cell, $\boldsymbol{\tau}_\mu$ the position of the μth ion within the unit cell, N the total number of the unit cells in a crystal, $\mathbf{k}$ a wave vector, and a specifies an atomic orbital. Then the matrix elements of the Hamiltonian are defined by

$$H_{\mu a \nu b}(\mathbf{k}, \mathbf{k}') = \langle \Phi_{\mu a \mathbf{k}} | \mathcal{H}_e | \Phi_{\nu b \mathbf{k}'} \rangle , \quad (13.10)$$

where $\mathcal{H}_e$ represents the effective one-electron Hamiltonian derived in Chap. 11. This Hamiltonian matrix is expressed by inserting (13.9) into (13.10) as follows;

$$H_{\mu a \nu b}(\boldsymbol{k},\boldsymbol{k}') = \sum_{l-l'} e^{-i\boldsymbol{k}\cdot\boldsymbol{R}_{l\mu}} e^{i\boldsymbol{k}'\cdot\boldsymbol{R}_{l'\nu}} H_{l\mu a, l'\nu b} , \quad (13.11)$$

where

$$H_{l\mu a, l'\nu b} = \langle \varphi_a(\boldsymbol{r} - \boldsymbol{R}_{l\mu} - \delta\boldsymbol{R}_{l\mu}) | \mathcal{H}_e | \varphi_b(\boldsymbol{r} - \boldsymbol{R}_{l'\nu} - \delta\boldsymbol{R}_{l'\nu}) \rangle . \quad (13.12)$$

The matrix element $H_{l\mu a,l'\nu b}$ is a function of $\boldsymbol{R}$ which is the difference between the position vectors of the two ions. In the following we calculate the many-body-effect included energy bands and expand them in terms of the atomic displacements $\delta\boldsymbol{R}_{l\mu}^\alpha$ or their Fourier transformations $u_{\boldsymbol{q}\mu}^\alpha$ defined by,

$$\delta\boldsymbol{R}_{l\mu}^\alpha = \frac{1}{\sqrt{N}} \sum_q e^{i\boldsymbol{q}\boldsymbol{R}_{l\mu}} u_{\boldsymbol{q}\mu}^\alpha , \quad (13.13)$$

where α indicates x, y and z. By expanding the energy bands up to the first order in $\delta\boldsymbol{R}_{l\mu}^\alpha$ or $u_{\boldsymbol{q}\mu}^\alpha$, the Hamiltonian matrix element $H_{\mu a \nu b}(\boldsymbol{k},\boldsymbol{k}')$ is expressed as

$$H_{\mu a \nu b}(\boldsymbol{k}\boldsymbol{k}') = H^0_{\mu a \nu b}(\boldsymbol{k})\delta_{\boldsymbol{k}\boldsymbol{k}'} + \sum_q \sum_{\mu'\alpha} \dot{T}^\alpha_{\mu'}(\mu a \boldsymbol{k}, \nu b \boldsymbol{k}') u_{\boldsymbol{q}\mu'}^\alpha \delta_{\boldsymbol{k}-\boldsymbol{q},\boldsymbol{k}'} . \quad (13.14)$$

Here $H^0_{\mu a \nu b}(\boldsymbol{k})$ is the Hamiltonian matrix element for an undistorted structure and $\dot{T}^\alpha_{\mu'}(\mu a \boldsymbol{k}, \nu b \boldsymbol{k}')$ is a quantity related to the derivative of a transfer interaction or of an on-site energy with regard to a displacement. The definition of $\dot{T}^\alpha_{\mu'}(\mu a \boldsymbol{k}, \nu b \boldsymbol{k}')$ is given as follows;

$$\dot{T}^\alpha_\mu(\mu' a \boldsymbol{k} \nu' b \boldsymbol{k}') = \frac{1}{\sqrt{N}}[\delta_{\mu\mu'} T^\alpha_{\mu' a,\ \nu' b}(\boldsymbol{k}') - \delta_{\mu\nu'} T^\alpha_{\mu' a,\ \nu' b}(\boldsymbol{k})] \quad \text{for } \mu' a \neq \nu' b ,$$

$$\dot{T}^\alpha_\mu(\mu' a \boldsymbol{k} \nu' b \boldsymbol{k}') = \frac{1}{\sqrt{N}} T^{\star\alpha}_{\mu c,\ \nu' b}(\boldsymbol{k}' - \boldsymbol{k}) \quad \text{for } \mu' a = \nu' b , \quad (13.15)$$

where

$$T^\alpha_{\mu' a,\ \nu' b}(\boldsymbol{k}) = \sum_{l-l'} e^{-i\boldsymbol{k}\cdot(\boldsymbol{R}_{l\mu'}-\boldsymbol{R}_{l'\nu'})} T^\alpha_{l\mu' a,\ l'\nu' b} , \quad (13.16)$$

$$T^{\star\alpha}_{\mu c,\ \nu' b}(\boldsymbol{k}) = \sum_{l-l'} e^{-i\boldsymbol{k}\cdot(\boldsymbol{R}_{l\mu}-\boldsymbol{R}_{l'\nu'})} T^{\star\alpha}_{l\mu c,\ l'\nu' b} , \quad (13.17)$$

$$T^\alpha_{l\mu' a,\ l'\nu' b} = \frac{\partial}{\partial R^\alpha} H_{l\mu' a,\ l'\nu' b}\Big|_{\boldsymbol{R}=\boldsymbol{R}_{l\mu'}-\boldsymbol{R}_{l'\nu'}} , \quad (13.18)$$

$$T^{\star\alpha}_{l\mu c,\ l'\mu' a}(\mu) = \frac{\partial}{\partial R^\alpha} H_{l'\mu' a,\ l'\mu' a}\Big|_{\boldsymbol{R}=\boldsymbol{R}_{l\mu}-\boldsymbol{R}_{l'\mu'}} . \quad (13.19)$$

Then the electron–phonon interaction in (13.3) is calculated on a tight binding model as follows;

$$V^\gamma(\boldsymbol{k}, \boldsymbol{k}-\boldsymbol{q}) = \sum_{\mu\alpha} \frac{1}{\sqrt{M_\mu}} \epsilon_{\gamma\mu\alpha}(\boldsymbol{q}) g_\mu^\alpha(\boldsymbol{k}\boldsymbol{k}-\boldsymbol{q}) \,, \qquad (13.20)$$

$$g_\mu^\alpha(\boldsymbol{k}\boldsymbol{k}') = \sum_{n\mu'a} \sum_{n'\nu'b} [A^\dagger(\boldsymbol{k})]_{n\mu'a} [\dot{T}_\mu^\alpha(\boldsymbol{k}\boldsymbol{k}')]_{\mu'a,\,\nu'b} [A(\boldsymbol{k}')]_{\nu'bn'} \, \mathcal{E}_\gamma \,, (13.21)$$

where

$$\begin{aligned}\mathcal{E}_\gamma &= 1 \quad \cdots \quad \text{for the process where pseudo-momentum } \boldsymbol{k}'-\boldsymbol{k}+\boldsymbol{q}=0 \\ &= -1 \quad \cdots \quad \text{for the process where pseudo-momentum } \boldsymbol{k}'-\boldsymbol{k}+\boldsymbol{q}=\boldsymbol{K}\,.\end{aligned}$$
(13.22)

In (13.20) and (13.21), $[A(\boldsymbol{k}')]_{\nu'bn'}$ is the ($\nu'bn'$)-th element of the transformation matrix in the undistorted structure, $\epsilon_{\gamma\mu\alpha}(\boldsymbol{q})$ the polarization vector of μth atom for a phonon mode γ with $\alpha = x, y, z$, and $\boldsymbol{K}$ the reciprocal lattice vector in the AF Brillouin zone. The detailed expressions of the electron–phonon matrix elements at the μ-th atom between $\boldsymbol{k}$ and $\boldsymbol{k}'$ states are given in the appendix at the end of this chapter.

In carrying out calculations of the spectral function for LSCO in the following chapter, one can see a reason why the electron–phonon interactions which scatter a pair of electrons from one pair state ($\boldsymbol{k}\uparrow, -\boldsymbol{k}\downarrow$) to a different pair state ($\boldsymbol{k}'\uparrow, -\boldsymbol{k}'\downarrow$) are repulsive for some combinations of ($\boldsymbol{k}, \boldsymbol{k}'$) while attractive for others for the K–S model.

13.3 Calculation of the Spectral Functions for s-, p- and d-waves

Following the method of Motizuki et al. [156, 157, 158], we will express the band structure numerically calculated in Chap. 11 in a tight binding analytical form, and calculate the spectral function $\alpha^2 F_{\uparrow\downarrow}(\Omega, \theta, \theta')$, by using the expressions of $g_\mu^\alpha(\boldsymbol{k}, \boldsymbol{k}')$ and $V^\gamma(\boldsymbol{k}, \boldsymbol{k}')$ based on the tight binding model, which are given in the appendix of this chapter. In the present theory, for the origins of the electron–phonon interactions $g_\mu^\alpha(\boldsymbol{k}, \boldsymbol{k}')$, we consider the change of both the transfer interactions and the on-site energies due to the displacement of atoms for each phonon mode [178, 196]. The change of the on-site energies has not been taken into account in the treatment of Motizuki et al. In the present theory, the derivatives of transfer integrals between Cu and O in CuO$_2$ plane are taken into account through the derivatives of the Slater Koster parameter, $t'_1(dp\sigma) = dt_1(dp\sigma)/dR = 2.6\,\mathrm{eV\mathring{A}^{-1}}$, calculated by DeWeert et al. [153]. (With regard to the Slater–Koster parameters, readers should read Chap. 10.) As for the effect of the displacement of atoms upon the on-site energies

144   13 Electron–Phonon Interaction and Electron–Phonon Spectral Functions

in the tight binding band, we calculate the change of the energies of the $^1A_{1g}$ and $^3B_{1g}$ multiplets, $dE_{A_{1g}}/dR$ and $dE_{B_{1g}}/dR$, by using the calculated results of energy difference with respect to the distance of Cu and apical O by Kamimura and Eto [104]. From the result of Kamimura and Eto we find that $E'_{B_{1g}} \equiv dE_{B_{1g}}/dR = 2.8\,\text{eV}\,\text{Å}^{-1}$ and $E'_{A_{1g}} \equiv dE_{A_{1g}}/dR = 2.2\,\text{eV}\,\text{Å}^{-1}$, where $E_{B_{1g}}$ and $E_{A_{1g}}$ denote the on-site energy of $^3B_{1g}$ and $^1A_{1g}$, respectively, as we already described in Chap. 5.

As an example of the calculated results, we present the calculated results of the θ and θ' dependence of the electron–phonon spectral functions $\alpha^2 F_{\uparrow\downarrow}(\Omega, \theta, \theta')$ for an A_{1g} phonon mode in Figs. 13.1–13.4, that for an E_u phonon mode in Figs. 13.5–13.8 and the calculated results of the θ and θ' dependence of $\alpha^2 F_{\uparrow\uparrow}(\Omega, \theta, \theta')$ for an A_{1g} phonon mode in Figs. 13.9–13.12 for $La_{2-x}Sr_xCuO_4$. For convenience of readers, all the figures from Fig. 13.1 to Fig. 13.16 are placed at the end of this section.

The θ-dependence of the spectral functions may be more easily understood from Fig. 13.13 and Fig. 13.14, where the calculated results of the spectral functions $\alpha^2 F_{\uparrow\downarrow}$ are shown, respectively, as a function of θ for fixed values of Ω and θ' for an A_{1g} phonon mode in which the apical oxygens move vertically

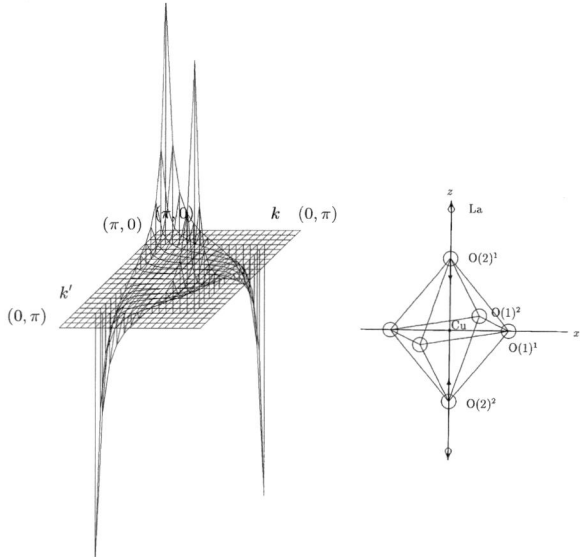

Fig. 13.1. The θ and θ' dependence of the momentum-dependent spectral function $\alpha^2 F_{\uparrow\downarrow}(\Omega, \theta, \theta')$ calculated for an A_{1g} phonon mode shown in the inset of the figure for fixed values of Ω and θ', in LSCO with tetragonal symmetry. The spectral function $\alpha^2 F_{\uparrow\downarrow}(\Omega, \theta, \theta')$ is shown for $0 \leq \theta \leq \pi/2$ and $0 \leq \theta' \leq \pi/2$. Here Cu–O–Cu distance a is taken as unity

13.3 Calculation of the Spectral Functions for s-, p- and d-waves   145

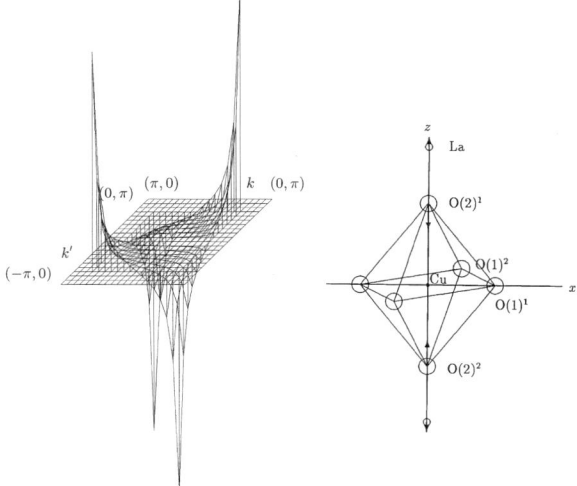

Fig. 13.2. The θ and θ' dependence of the momentum-dependent spectral function $\alpha^2 F_{\uparrow\downarrow}(\Omega, \theta, \theta')$ calculated for an A_{1g} phonon mode shown in the inset of the figure for fixed values of Ω and θ', in LSCO with tetragonal symmetry. The spectral function $\alpha^2 F_{\uparrow\downarrow}(\Omega, \theta, \theta')$ is shown for $\pi/2 \leq \theta \leq \pi$ and $0 \leq \theta' \leq \pi/2$. Here Cu–O–Cu distance a is taken as unity

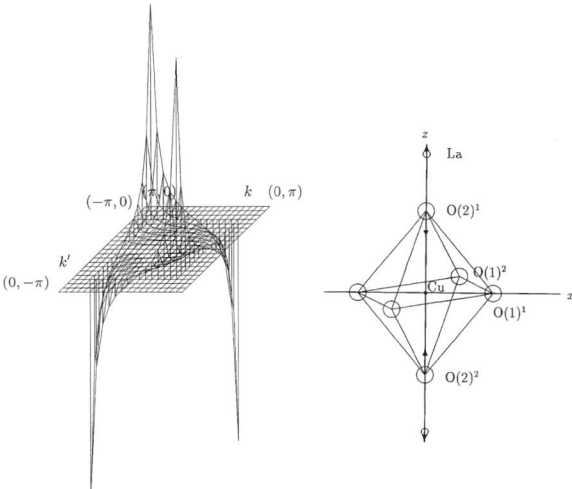

Fig. 13.3. The θ and θ' dependence of the momentum-dependent spectral function $\alpha^2 F_{\uparrow\downarrow}(\Omega, \theta, \theta')$ calculated for an A_{1g} phonon mode shown in the inset of the figure for fixed values of Ω and θ', in LSCO with tetragonal symmetry. The spectral function $\alpha^2 F_{\uparrow\downarrow}(\Omega, \theta, \theta')$ is shown for $\pi \leq \theta \leq 3\pi/2$ and $0 \leq \theta' \leq \pi/2$. Here Cu–O–Cu distance a is taken as unity

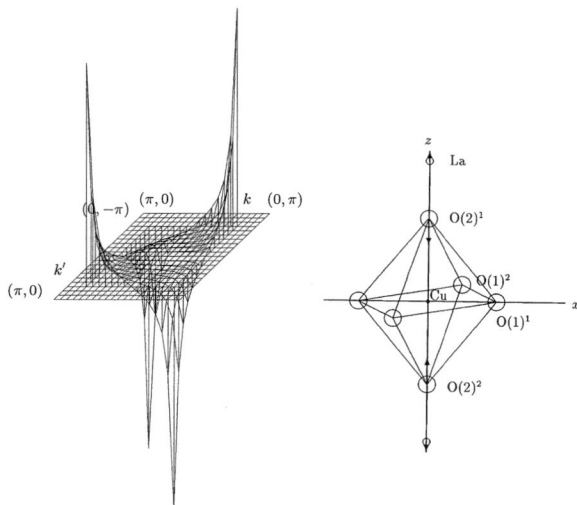

Fig. 13.4. The θ and θ' dependence of the momentum-dependent spectral function $\alpha^2 F_{\uparrow\downarrow}(\Omega, \theta, \theta')$ calculated for an A_{1g} phonon mode shown in the inset of the figure for fixed values of Ω and θ', in LSCO with tetragonal symmetry. The spectral function $\alpha^2 F_{\uparrow\downarrow}(\Omega, \theta, \theta')$ is shown for $3\pi/2 \leq \theta \leq 2\pi$ and $0 \leq \theta' \leq \pi/2$. Here Cu–O–Cu distance a is taken as unity

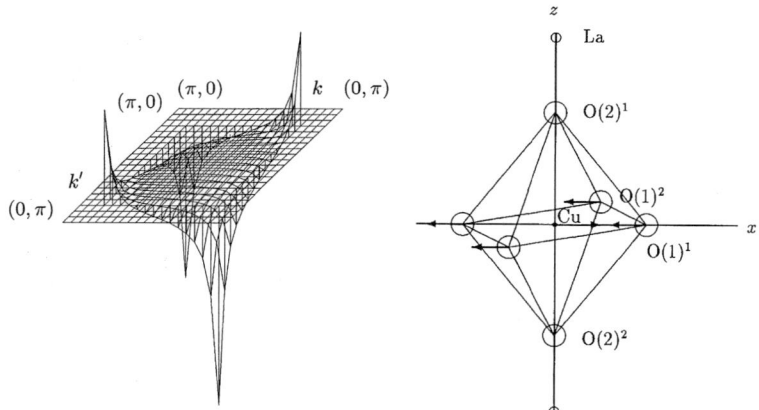

Fig. 13.5. The θ and θ' dependence of the momentum-dependent spectral function $\alpha^2 F_{\uparrow\downarrow}(\Omega, \theta, \theta')$ calculated for an E_u phonon mode shown in the inset of the figure. The spectral function $\alpha^2 F_{\uparrow\downarrow}(\Omega, \theta, \theta')$ is shown for fixed values of Ω and θ', in LSCO with tetragonal symmetry. The spectral function $\alpha^2 F_{\uparrow\downarrow}(\Omega, \theta, \theta')$ is shown for $0 \leq \theta \leq \pi/2$ and $0 \leq \theta' \leq \pi/2$. Here Cu–O–Cu distance a is taken as unity

13.3 Calculation of the Spectral Functions for s-, p- and d-waves   147

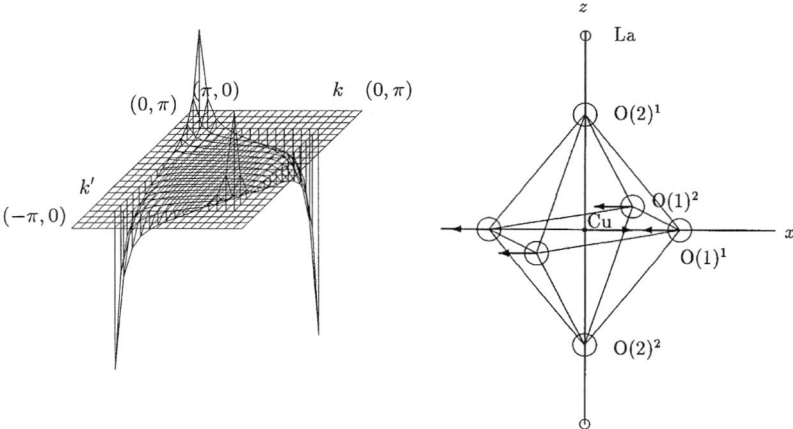

Fig. 13.6. The θ and θ' dependence of the momentum-dependent spectral function $\alpha^2 F_{\uparrow\downarrow}(\Omega,\theta,\theta')$ calculated for an E_u phonon mode shown in the inset of the figure. The spectral function $\alpha^2 F_{\uparrow\downarrow}(\Omega,\theta,\theta')$ is shown for fixed values of Ω and θ', in LSCO with tetragonal symmetry. The spectral function $\alpha^2 F_{\uparrow\downarrow}(\Omega,\theta,\theta')$ is shown for $0 \leq \theta \leq \pi/2$ and $0 \leq \theta' \leq \pi/2$. Here Cu–O–Cu distance a is taken as unity

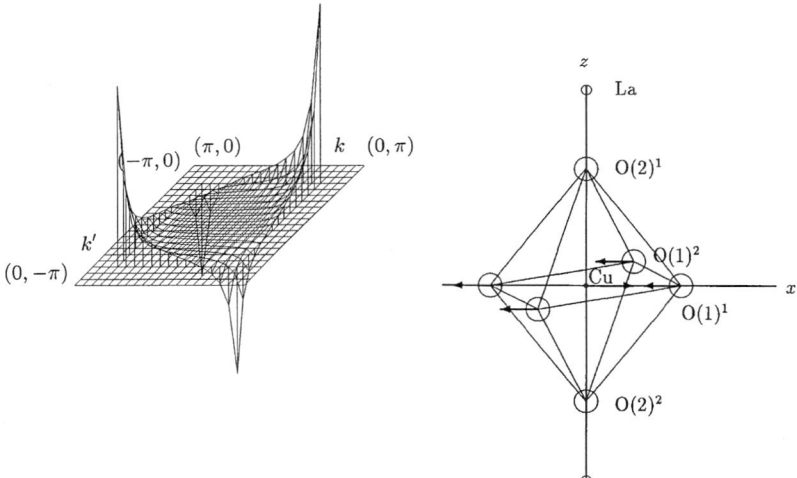

Fig. 13.7. The θ and θ' dependence of the momentum-dependent spectral function $\alpha^2 F_{\uparrow\downarrow}(\Omega,\theta,\theta')$ calculated for an E_u phonon mode shown in the inset of the figure. The spectral function $\alpha^2 F_{\uparrow\downarrow}(\Omega,\theta,\theta')$ is shown for fixed values of Ω and θ', in LSCO with tetragonal symmetry. The spectral function $\alpha^2 F_{\uparrow\downarrow}(\Omega,\theta,\theta')$ is shown for $0 \leq \theta \leq \pi/2$ and $0 \leq \theta' \leq \pi/2$. Here Cu–O–Cu distance a is taken as unity

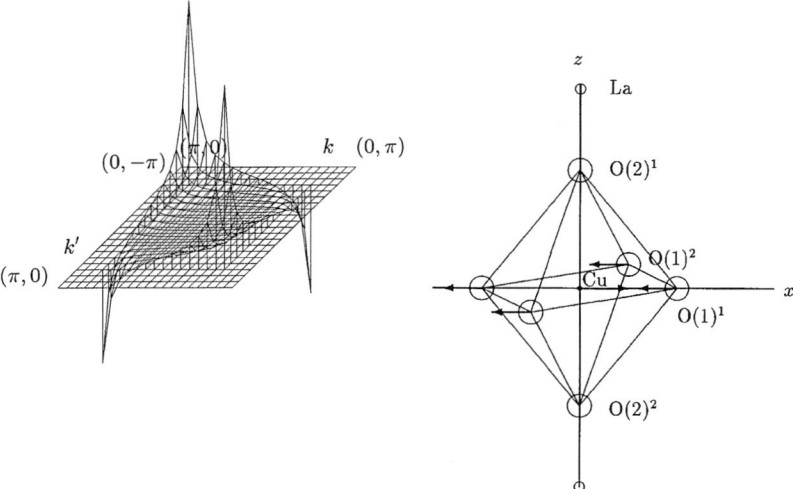

Fig. 13.8. The θ and θ' dependence of the momentum-dependent spectral function $\alpha^2 F_{\uparrow\downarrow}(\Omega,\theta,\theta')$ calculated for an E_u phonon mode shown in the inset of the figure for fixed values of Ω and θ', in LSCO with tetragonal symmetry. The spectral function $\alpha^2 F_{\uparrow\downarrow}(\Omega,\theta,\theta')$ is shown for $0 \leq \theta \leq \pi/2$ and $0 \leq \theta' \leq \pi/2$. Here Cu–O–Cu distance a is taken as unity

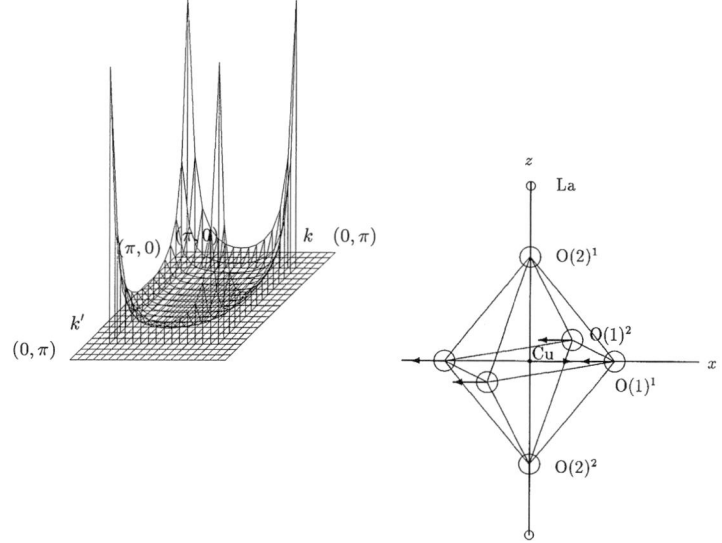

Fig. 13.9. The θ and θ' dependence of the momentum-dependent spectral function $\alpha^2 F_{\uparrow\uparrow}(\Omega,\theta,\theta')$ calculated for an E_u phonon mode shown in the inset of this figure, for fixed values of Ω. The spectral function $\alpha^2 F_{\uparrow\uparrow}(\Omega,\theta,\theta')$ is shown for $0 \leq \theta \leq \pi/2$ and $0 \leq \theta' \leq \pi/2$. Here Cu–O–Cu distance a is taken as unity

13.3 Calculation of the Spectral Functions for s-, p- and d-waves   149

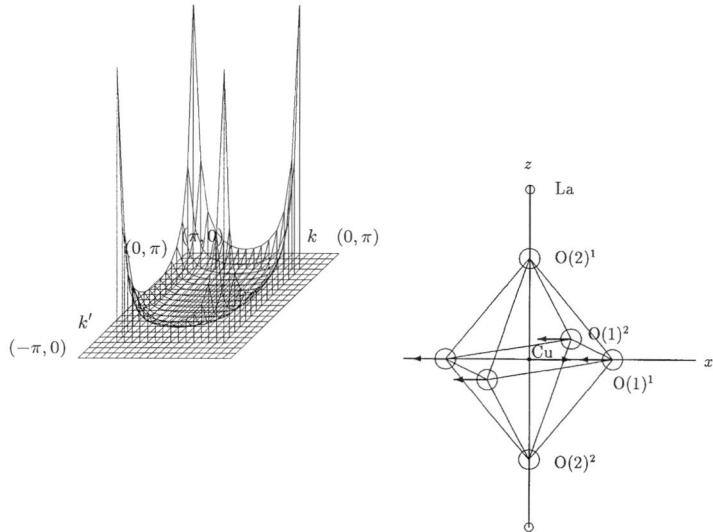

Fig. 13.10. The θ and θ' dependence of the momentum-dependent spectral function $\alpha^2 F_{\uparrow\uparrow}(\Omega,\theta,\theta')$ calculated for an E_u phonon mode shown in the inset of this figure, for fixed values of Ω. The spectral function $\alpha^2 F_{\uparrow\uparrow}(\Omega,\theta,\theta')$ is shown for $\pi/2 \leq \theta \leq \pi$ and $0 \leq \theta' \leq \pi/2$. Here Cu–O–Cu distance a is taken as unity

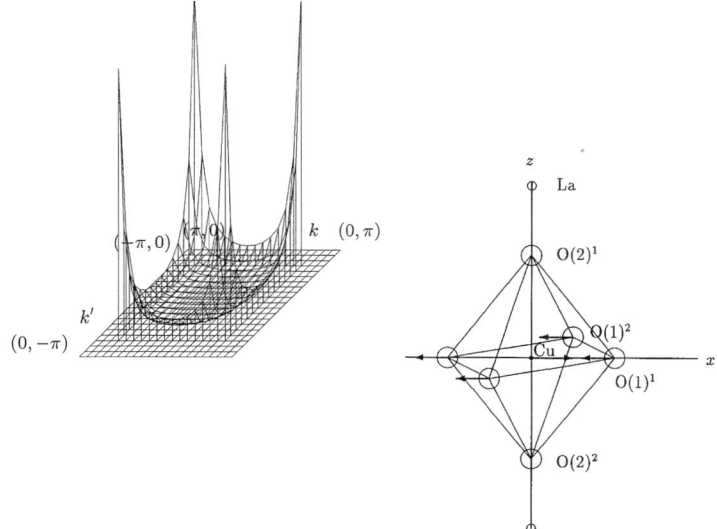

Fig. 13.11. The θ and θ' dependence of the momentum-dependent spectral function $\alpha^2 F_{\uparrow\uparrow}(\Omega,\theta,\theta')$ calculated for an E_u phonon mode shown in the inset of this figure, for fixed values of Ω. The spectral function $\alpha^2 F_{\uparrow\uparrow}(\Omega,\theta,\theta')$ is shown for $\pi \leq \theta \leq 3\pi/2$ and $0 \leq \theta' \leq \pi/2$. Here Cu–O–Cu distance a is taken as unity

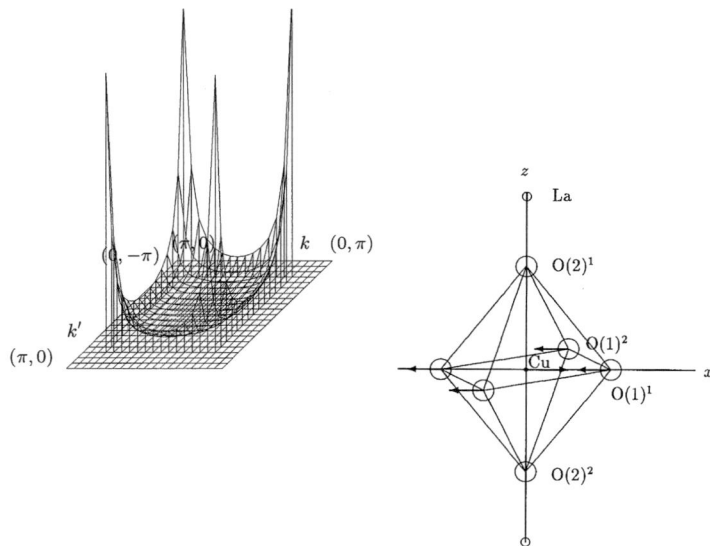

Fig. 13.12. The θ and θ' dependence of the momentum-dependent spectral function $\alpha^2 F_{\uparrow\uparrow}(\Omega,\theta,\theta')$ calculated for an E_u phonon mode shown in the inset of this figure, for fixed values of Ω. The spectral function $\alpha^2 F_{\uparrow\uparrow}(\Omega,\theta,\theta')$ is shown for $3\pi/2 \leq \theta \leq 2\pi$ and $0 \leq \theta' \leq \pi/2$. Here Cu–O–Cu distance a is taken as unity

for a CuO$_2$ plane (Fig. 13.13) and for an E_u phonon mode in which the oxygen ions move within a CuO$_2$ plane (Fig. 13.14).

As seen in Figs. 13.1–13.8, the momentum-dependent spectral functions for a singlet Cooper pair, $\alpha^2 F_{\uparrow\downarrow}(\Omega,\theta,\theta')$, shows a sharp $\boldsymbol{k}$-dependence. Its $\boldsymbol{k}$-dependence follows d$_{x^2-y^2}$ symmetry. The sharp peaks of the spectral function near G$_1$ points, $(\pm\pi/a, 0, 0)$ and $(0, \pm\pi/a, 0)$, are due to the appearance of the van Hove singularity in the density of states (DOS) at G$_1$ points. Summarizing the calculated results of $\alpha^2 F_{\uparrow\downarrow}(\Omega,\theta,\theta')$ for LSCO, we can say that the s-component almost vanishes while the d-component is conspicuous.

Then, by summing up the contributions from all the phonon modes shown in Table 13.1 at the end of this chapter to the spectral function, we have calculated the s-wave component of spectral function $\alpha^2 F_{\uparrow\downarrow}^{(0)}(\Omega)$ and its d-wave component $\alpha^2 F_{\uparrow\downarrow}^{(2)}(\Omega)$, which are defined from $\alpha^2 F_{\uparrow\downarrow}(\Omega,\theta,\theta')$ as follows:

$$\alpha^2 F_{\uparrow\downarrow}(\Omega,\theta,\theta') = \frac{1}{2\pi} \sum_{n=0}^{\infty} \alpha^2 F_{\uparrow\downarrow}^{(n)}(\Omega) \cos n\theta \cos n\theta' . \qquad (13.23)$$

As will be explained in the next chapter, the d-wave component $\alpha^2 F_{\uparrow\downarrow}^{(2)}(\Omega)$ contributes to the appearance of d-wave superconductivity. As a result we find that, for phonon modes in which oxygen and copper ions vibrate within a CuO$_2$ plane such as a breathing mode, the d-wave component in the spectral

13.3 Calculation of the Spectral Functions for s-, p- and d-waves

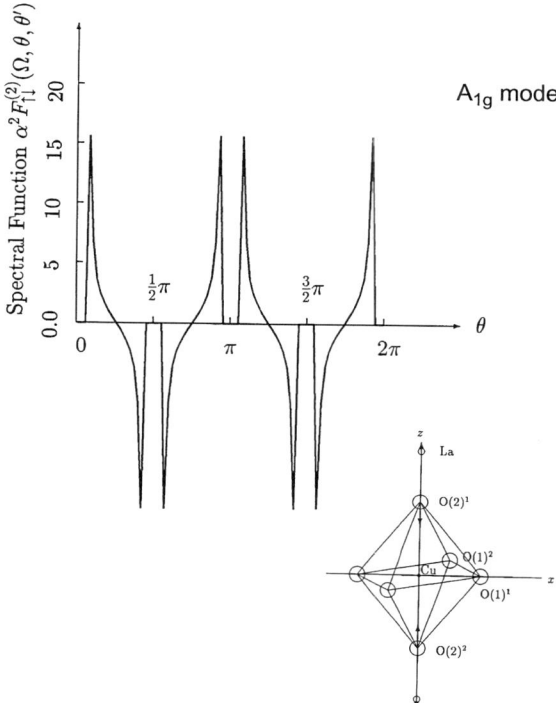

Fig. 13.13. The θ dependence of the momentum-dependent spectral function $\alpha^2 F_{\uparrow\downarrow}^{(2)}(\Omega, \theta, \theta')$ calculated for an A_{1g} phonon mode shown at the *bottom* of this figure for fixed values of Ω and θ'. In this case θ' is taken a value near G_1 point, i.e., $(\pi/a, 0, 0)$. Here Cu–O–Cu distance a is taken as unity

function is negative (repulsive) and the s-wave component is very small. Thus the in-plane modes do not contribute to the formation of Cooper pairs. On the other hand, for the phonon modes in which the apical oxygen ions and La ions move vertically for CuO_2 plane, the d-wave component of the spectral function, $\alpha^2 F_{\uparrow\downarrow}^{(2)}(\Omega)$, has a positive (attractive) sign, and the s wave component is very small. In Fig. 13.15 we show the calculated results of the total contribution for the d-wave components of the spectral function, $\alpha^2 F_{\uparrow\downarrow}^{(2)}(\Omega)$, from all the phonon modes of LSCO.

We can obtain the value of electron–phonon coupling constant for d-wave pairing, λ_d by integrating the positive part of the calculated d-wave components of the spectral function $\alpha^2 F_{\uparrow\downarrow}^{(2)}(\Omega)$ over the phonon frequency Ω, as shown in the following expression.

$$\lambda_d = 2 \int_0^\infty \frac{\frac{1}{2}\alpha^2 F_{\uparrow\downarrow}^{(2)}(\Omega)}{\Omega} d\Omega . \quad (13.24)$$

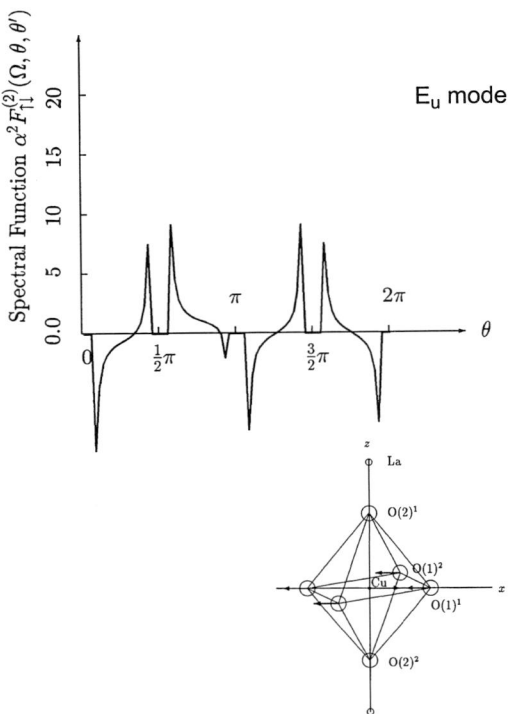

Fig. 13.14. The θ dependence of the momentum-dependent spectral function $\alpha^2 F_{\uparrow\downarrow}(\Omega, \theta, \theta')$ calculated for an E$_u$ phonon mode shown at the *bottom* of this figure for fixed values of Ω and θ'. In this case θ' is taken a value near G$_1$ point, i.e., $(\pi/a, 0, 0)$. Here Cu–O–Cu distance a is taken as unity

The value of λ_d thus calculated for $x = 0.15$ is 1.96. From this result we can say that LSCO is the superconductor of a strong coupling. In Fig. 13.16 we also give the calculated results of the total contribution to the spectral function $\alpha^2 F_{\uparrow\uparrow}(\Omega)$. In this case only the s-component $\alpha^2 F_{\uparrow\uparrow}^{(0)}(\Omega)$ appears. This causes a mass enhancement of the electronic states.

In the electron– and spin–structures of the K–S model characterized by the alternant appearance of the a$_{1g}^*$ and the b$_{1g}$ orbitals and by the different spatial distribution of Bloch wave functions for up-spin and down-spin dopant holes, the electron–phonon interaction matrix element for an up-spin hole-carrier, $V_\uparrow^\gamma(\boldsymbol{k}, \boldsymbol{k}')$, becomes different from that for a down-spin carrier, $V_\downarrow^\gamma(\boldsymbol{k}, \boldsymbol{k}')$ and thus, $\alpha^2 F_{\uparrow\downarrow}(\Omega, \theta, \theta')$ changes its sign. This situation makes the $\boldsymbol{k}$ dependence of $\alpha^2 F_{\uparrow\downarrow}(\Omega, \theta, \theta')$ much more conspicuous and stronger than in the case of an ordinary BCS case. We can say that the present results of $\alpha^2 F_{\uparrow\uparrow}^{(2)}(\Omega)$ are probably the first quantitative calculation of the electron–phonon spectral function for a d-wave superconductivity in real materials.

13.3 Calculation of the Spectral Functions for s-, p- and d-waves 153

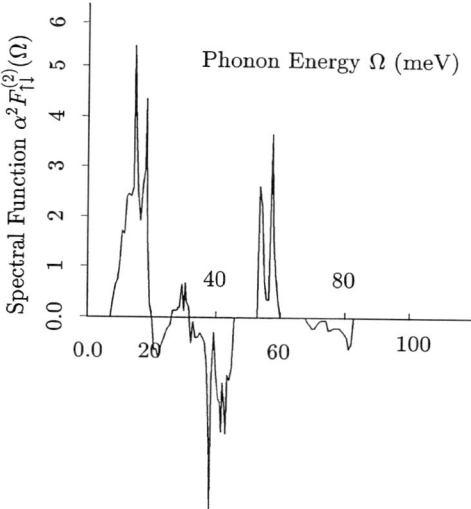

Fig. 13.15. The d-wave component of the spectral function, $\alpha^2 F^{(2)}_{\uparrow\downarrow}(\Omega)$, calculated for LSCO with $x = 0.15$, due to the contribution from all the phonon modes in LSCO

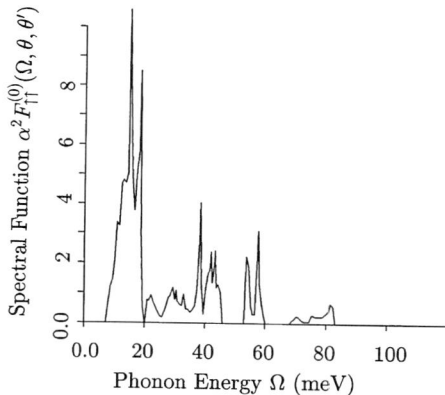

Fig. 13.16. The spectral function, which contributes to mass enhancement, $\alpha^2 F^{(0)}_{\uparrow\uparrow}(\Omega)$, calculated for LSCO with $x = 0.15$, due to the contribution from all the phonon modes in LSCO

Table 13.1. Normal modes corresponding to Δ-line, $(0,0)\to(\pi,0)$, for LSCO. The mass ratio to satisfy the orthogonality relation are omitted in the table. See S. Mase et al., Phonon Dispersion Curves of High T_c Superconductors I. $(La_{1-x}Sr_x)_2CuO_4$, J. Phys. Soc. Jpn. **57** (1988) 607

		La^1	La^2	Cu	$O(1)^1$	$O(1)^2$	$O(2)^1$	$O(2)^2$	mode
La		(100)	(100)	(100)	(100)	(100)	(100)	(100)	E_u
		(010)	(010)	(010)	(010)	(010)	(010)	(010)	E_u
		(001)	(001)	(001)	(001)	(001)	(001)	(001)	A_{2u}
		(100)	($\bar{1}$00)				(100)	($\bar{1}$00)	E_g
		(010)	(0$\bar{1}$0)				(010)	(0$\bar{1}$0)	E_g
		(001)	(00$\bar{1}$)				(001)	(00$\bar{1}$)	A_{1g}
Cu		($\frac{3}{4}$00)	($\frac{3}{4}$00)	(100)	(100)	(100)	($\frac{3}{4}$00)	($\frac{3}{4}$00)	E_u
		(0$\frac{3}{4}$0)	(0$\frac{3}{4}$0)	(010)	(010)	(010)	(0$\frac{3}{4}$0)	(0$\frac{3}{4}$0)	E_u
		(00$\frac{3}{4}$)	(00$\frac{3}{4}$)	(001)	(001)	(001)	(00$\frac{3}{4}$)	(00$\frac{3}{4}$)	A_{2u}
O(1) s					(100)	($\bar{1}$00)			E_u
					(010)	(0$\bar{1}$0)			E_u
O(1) b				(200)	(100)	(100)			E_u
				(020)	(010)	(010)			E_u
				(002)	(001)	(001)			A_{2u}
					(001)	(00$\bar{1}$)			B_{2u}
O(2) s	(001)	(001)					(00$\bar{1}$)	(00$\bar{1}$)	A_{2u}
	(001)	(001)					(001)	(001)	A_{1g}
O(2) b	(100)	(100)					($\bar{1}$00)	($\bar{1}$00)	E_u
	(010)	(010)					(0$\bar{1}$0)	(0$\bar{1}$0)	E_u
	(100)	(100)					($\bar{1}$00)	($\bar{1}$00)	E_g
	(010)	(010)					(0$\bar{1}$0)	(0$\bar{1}$0)	E_g

In the following chapter we will show that this characteristic k dependence produces a large d-wave component of the spectral function. In the ordinary BCS case, a d-wave component is always small because of the positive definite k dependence of $\alpha^2 F_{\uparrow\downarrow}(\Omega,\theta,\theta')$. Thus the present electron– and spin–structures in the K–S model are the key factors in creating d-wave pairing in the phonon mechanism [30].

In the following chapter we will calculate various superconducting properties of cuprates, such as the hole-concentration dependence of T_c, isotope effects, etc.

13.4 Appendix

Appendix D. The Explicit Forms of the Electron–Phonon Interaction

In this appendix the explicit forms of the electron–phonon interaction $g_\mu^\alpha(\boldsymbol{k}\boldsymbol{k}')$ are given for $\mu = $ Cu, $O(1)^1$, $O(1)^2$, $O(2)^1$ and $O(2)^2$ which are the copper

13.4 Appendix

atom, the oxygen atoms in CuO$_2$ plane and the apical oxygen atoms, respectively.

$$g^x_{\text{Cu}}(\boldsymbol{k}\boldsymbol{k}') = -\frac{1}{\sqrt{N}} [\, A_s^*(\boldsymbol{k})_{16} A(\boldsymbol{k}')_1 T^x_{1,16}(\boldsymbol{k}')^* + A^*(\boldsymbol{k})_1 A_s(\boldsymbol{k}')_{16} T^x_{1,16}(\boldsymbol{k})$$
$$+ A_s^*(\boldsymbol{k})_{17} A(\boldsymbol{k}')_1 T^x_{1,17}(\boldsymbol{k}')^* + A^*(\boldsymbol{k})_1 A_s(\boldsymbol{k}')_{17} T^x_{1,17}(\boldsymbol{k})\,]\,,$$

$$g^x_{\text{O}(1)1}(\boldsymbol{k}\boldsymbol{k}') = +\frac{1}{\sqrt{N}} [\, A_s^*(\boldsymbol{k})_1 A(\boldsymbol{k}')_{16} T^x_{1,16}(\boldsymbol{k}') + A^*(\boldsymbol{k})_{16} A_s(\boldsymbol{k}')_1 T^x_{1,16}(\boldsymbol{k})^*$$
$$+ A_s^*(\boldsymbol{k})_1 A(\boldsymbol{k}')_{17} T^x_{1,17}(\boldsymbol{k}') + A^*(\boldsymbol{k})_{17} A_s(\boldsymbol{k}')_1 T^x_{1,17}(\boldsymbol{k})^*\,]\,,$$

$$g^y_{\text{Cu}}(\boldsymbol{k}\boldsymbol{k}') = -\frac{1}{\sqrt{N}} [\, A_s^*(\boldsymbol{k})_{16} A(\boldsymbol{k}')_5 T^y_{5,16}(\boldsymbol{k}')^* + A^*(\boldsymbol{k})_5 A_s(\boldsymbol{k}')_{16} T^y_{5,16}(\boldsymbol{k})$$
$$+ A_s^*(\boldsymbol{k})_{17} A(\boldsymbol{k}')_5 T^y_{5,17}(\boldsymbol{k}')^* + A^*(\boldsymbol{k})_5 A_s(\boldsymbol{k}')_{17} T^y_{5,17}(\boldsymbol{k})\,]\,,$$

$$g^y_{\text{O}(1)2}(\boldsymbol{k}\boldsymbol{k}') = +\frac{1}{\sqrt{N}} [\, A_s^*(\boldsymbol{k})_5 A(\boldsymbol{k}')_{16} T^y_{5,16}(\boldsymbol{k}') + A^*(\boldsymbol{k})_{16} A_s(\boldsymbol{k}')_5 T^y_{5,16}(\boldsymbol{k})^*$$
$$+ A_s^*(\boldsymbol{k})_5 A(\boldsymbol{k}')_{17} T^y_{5,17}(\boldsymbol{k}') + A^*(\boldsymbol{k})_{17} A_s(\boldsymbol{k}')_5 T^y_{5,17}(\boldsymbol{k})^*\,]\,,$$

$$g^z_{\text{Cu}}(\boldsymbol{k}\boldsymbol{k}') = -\frac{1}{\sqrt{N}} [\, A_s^*(\boldsymbol{k})_{17} A(\boldsymbol{k}')_9 T^z_{9,17}(\boldsymbol{k}')^* + A^*(\boldsymbol{k})_9 A_s(\boldsymbol{k}')_{17} T^z_{9,17}(\boldsymbol{k})$$
$$+ A_s^*(\boldsymbol{k})_{17} A(\boldsymbol{k}')_{12} T^z_{12,17}(\boldsymbol{k}')^* + A^*(\boldsymbol{k})_{12} A_s(\boldsymbol{k}')_{17} T^z_{12,17}(\boldsymbol{k})\,]\,,$$

$$g^z_{\text{O}(2)1}(\boldsymbol{k}\boldsymbol{k}') = +\frac{1}{\sqrt{N}} [\, A_s^*(\boldsymbol{k})_9 A(\boldsymbol{k}')_{17} T^z_{9,17}(\boldsymbol{k}') + A^*(\boldsymbol{k})_{17} A_s(\boldsymbol{k}')_9 T^z_{9,17}(\boldsymbol{k})^*$$
$$+ A_s^*(\boldsymbol{k})_{17} A(\boldsymbol{k}')_{17} T^{*z}_{9,17}(\boldsymbol{k}'-\boldsymbol{k}) + A_s^*(\boldsymbol{k})_{16} A(\boldsymbol{k}')_{16} T^{*z}_{9,16}(\boldsymbol{k}'-\boldsymbol{k})]\,,$$

$$g^z_{\text{O}(2)2}(\boldsymbol{k}\boldsymbol{k}') = +\frac{1}{\sqrt{N}} [\, A_s^*(\boldsymbol{k})_{12} A(\boldsymbol{k}')_{17} T^z_{12,17}(\boldsymbol{k}') + A^*(\boldsymbol{k})_{17} A_s(\boldsymbol{k}')_{12} T^z_{12,17}(\boldsymbol{k})^*$$
$$+ A_s^*(\boldsymbol{k})_{17} A(\boldsymbol{k}')_{17} T^{*z}_{12,17}(\boldsymbol{k}'-\boldsymbol{k}) + A_s^*(\boldsymbol{k})_{16} A(\boldsymbol{k}')_{16} T^{*z}_{12,16}(\boldsymbol{k}'-\boldsymbol{k})]\,,$$

(13.25)

where $A(\boldsymbol{k})_i$ is the ith element of the transformation matrix and $A_s(\boldsymbol{k})_i$ is defined as,

$$A_s(\boldsymbol{k})_i = -A(\boldsymbol{k})_i \quad \text{when} \quad \gamma \text{ is 2nd-mode and } i \text{ is } B\text{-site}$$
$$= A(\boldsymbol{k})_i \quad \text{otherwise}\,.$$

(13.26)

The matrix elements of the transfer interactions in (8.10), between an A-site and another A-site or between a B-site and another B-site, $T^\alpha_{\mu' a \nu' b}(\boldsymbol{k})$, are expressed as follows,

$$T^x_{1,16}(\boldsymbol{k}) = T^x_{\text{O}(1)1x\ \text{Cu}x^2-y^2}(\boldsymbol{k}) = \frac{\sqrt{3}}{2} t'_1(pd\sigma) e^{-i\frac{k_x a}{2}} = -T^x_{16,1}(\boldsymbol{k})^*\,,$$

$$T^x_{1,17}(\boldsymbol{k}) = T^x_{\text{O}(1)1x\ \text{Cu}z^2}(\boldsymbol{k}) = -\frac{1}{2} t'_1(pd\sigma) e^{-i\frac{k_x a}{2}} = -T^x_{17,1}(\boldsymbol{k})^*\,,$$

$$T^y_{5,16}(\boldsymbol{k}) = T^y_{\text{O}(1)2y\ \text{Cu}x^2-y^2}(\boldsymbol{k}) = -\frac{\sqrt{3}}{2} t'_1(pd\sigma) e^{-i\frac{k_y a}{2}} = -T^y_{16,5}(\boldsymbol{k})^*\,,$$

$$T^y_{5,17}(\boldsymbol{k}) = T^y_{\text{O}(1)2y\ \text{Cu}z^2}(\boldsymbol{k}) = -\frac{1}{2} t'_1(pd\sigma) e^{-i\frac{k_y a}{2}} = -T^y_{17,5}(\boldsymbol{k})^*\,,$$

$$T^z_{9,17}(\boldsymbol{k}) = T^z_{\text{O}(2)1z\ \text{Cu}z^2}(\boldsymbol{k}) = t'_2(pd\sigma) e^{-i\, k_z 0.364 c/2} = -T^z_{17,9}{}^*(\boldsymbol{k})\,,$$

$$T^z_{12,17}(\boldsymbol{k}) = T^z_{\text{O}(2)2z\ \text{Cu}z^2}(\boldsymbol{k}) = t'_2(pd\sigma) e^{i\, k_z 0.364 c/2} = -T^z_{17,12}{}^*(\boldsymbol{k})\,,$$

$$T^{\star z}_{9,16}(\boldsymbol{k}) = T^{\star z}_{\mathrm{O}(2)^1 z\ \mathrm{Cu} x^2-y^2}(\boldsymbol{k}) = E'_{\mathrm{A_{1g}}} e^{-i\ k_z 0.364 c/2} = T^{\star z *}_{12,16}(\boldsymbol{k})\ ,$$
$$T^{\star z}_{9,17}(\boldsymbol{k}) = T^{\star z}_{\mathrm{O}(2)^2 z\ \mathrm{Cu} z^2}(\boldsymbol{k}) = E'_{\mathrm{B_{1g}}} e^{-i\ k_z 0.364 c/2} T^{\star z *}_{12,17}(\boldsymbol{k})$$
(13.27)

where $t'_i(pd\sigma)$ is the derivative of transfer integral between a Cu d orbital and a neighbouring O p orbital. The transfer integral $t_i(pd\sigma)$ have been defined in Sect. 3.2. For the matrix elements between A-site and B-site, those are expressed as

$$\begin{aligned}
T^x_{1,16}(\boldsymbol{k}) &= T^x_{\mathrm{O}(1)^1 x\ \mathrm{Cu} x^2-y^2}(\boldsymbol{k}) = \frac{\sqrt{3}}{2} t'_1(pd\sigma) e^{i\frac{k_x a}{2}} &= -T^x_{16,1}(\boldsymbol{k})^*\ , \\
T^x_{1,17}(\boldsymbol{k}) &= T^x_{\mathrm{O}(1)^1 x\ \mathrm{Cu} z^2}(\boldsymbol{k}) = -\frac{1}{2} t'_1(pd\sigma) e^{i\frac{k_x a}{2}} &= -T^x_{17,1}(\boldsymbol{k})^*\ , \\
T^y_{5,16}(\boldsymbol{k}) &= T^y_{\mathrm{O}(1)^2 y\ \mathrm{Cu} x^2-y^2}(\boldsymbol{k}) = -\frac{\sqrt{3}}{2} t'_1(pd\sigma) e^{i\frac{k_y a}{2}} &= -T^y_{16,5}(\boldsymbol{k})^*\ , \\
T^y_{5,17}(\boldsymbol{k}) &= T^y_{\mathrm{O}(1)^2 y\ \mathrm{Cu} z^2}(\boldsymbol{k}) = -\frac{1}{2} t'_1(pd\sigma) e^{i\frac{k_y a}{2}} &= -T^y_{17,5}(\boldsymbol{k})^*\ , \\
T^z_{9,17}(\boldsymbol{k}) &= T^z_{\mathrm{O}(2)^1 x\ \mathrm{Cu} z^2}(\boldsymbol{k}) = 0 &= -T^z_{17,9}{}^*(\boldsymbol{k})\ , \\
T^z_{12,17}(\boldsymbol{k}) &= T^z_{\mathrm{O}(2)^2 x\ \mathrm{Cu} z^2}(\boldsymbol{k}) = 0 &= -T^z_{17,12}{}^*(\boldsymbol{k})\ .
\end{aligned}$$
(13.28)

Appendix E. Repulsive Electron–Phonon Interaction between Up- and Down-Spin Carriers with Different Wave Function

As we mentioned in Chap. 8, the wave function for up-spin carriers is different from that for down-spin carriers, which makes the electron–phonon coupling constant for up-spin carriers different from that for down-spin carriers. Now we will explain why $V^\gamma_\uparrow(\boldsymbol{k},\boldsymbol{k}')$ is different from $V^\gamma_\downarrow(\boldsymbol{k},\boldsymbol{k}')$. Any phonon mode in the ordinary Brillouin zone develops two branches, as a result of folding it into the AF Brillouin zone. One branch corresponds to "acoustic type mode" in which the motion of the two neighbouring CuO_6 octahedra with localized up- and down-spins is the same except for the phase factor $\exp(i\boldsymbol{q}\cdot\boldsymbol{a})$, while the other corresponds to "optic type mode" in which the motion of the two neighbouring CuO_6 octahedra is opposite except for the phase factor $\exp(i\boldsymbol{q}\cdot\boldsymbol{a})$. If we denote the positions of two atoms in the lth AF unit cell whose distance is separated by $\boldsymbol{a}$, the translation vector from one Cu atom to a neighbouring Cu atom, by $\boldsymbol{R}_{l\mu 1}$ and $\boldsymbol{R}_{l\mu 2}$, then $\boldsymbol{R}_{l\mu 2} = \boldsymbol{R}_{l\mu 1} + \boldsymbol{a}$. The displacement of the atom at $\boldsymbol{R}_{l\mu 2}$, $\delta\boldsymbol{R}_{l\mu 2}$, is related to that at $\boldsymbol{R}_{l\mu 1}$, $\delta\boldsymbol{R}_{l\mu 1}$, as

$$\delta\boldsymbol{R}_{l\mu 2} = \pm \exp(i\boldsymbol{q}\boldsymbol{a}) \delta\boldsymbol{R}_{l\mu 1}\ ,$$
(13.29)

where the sign + and − correspond to "acoustic type phonon mode" and "optic type phonon mode" respectively. Here it should be noted that $\boldsymbol{a}$ is a non-primitive translation vector in the AF unit cell, though it is a primitive translation vector in the ordinary unit cell. By using Bloch theorem, the wave functions for up- and down-spin carriers are written as

$$\Psi_{\boldsymbol{k}\uparrow}(\boldsymbol{r}) = \exp(i\boldsymbol{k}\boldsymbol{r})u_{\boldsymbol{k}\uparrow}(\boldsymbol{r}) \tag{13.30}$$

$$\Psi_{\boldsymbol{k}\downarrow}(\boldsymbol{r}) = \exp(i\boldsymbol{k}\boldsymbol{r})u_{\boldsymbol{k}\downarrow}(\boldsymbol{r}) \tag{13.31}$$

where $u_{\boldsymbol{k}\uparrow}(\boldsymbol{r})$ and $u_{\boldsymbol{k}\downarrow}(\boldsymbol{r})$ have the periodicity of the lattice of the AF unit cell. In the present model, the effective Hamiltonian for up- and down-spin carriers, $H_{\text{eff}\uparrow}(\boldsymbol{r})$ and $H_{\text{eff}\downarrow}(\boldsymbol{r})$, satisfy the relation $H_{\text{eff}\downarrow}(\boldsymbol{r}+\boldsymbol{a}) = H_{\text{eff}\uparrow}(\boldsymbol{r})$, and $u_{\boldsymbol{k}\uparrow}(\boldsymbol{r})$ and $u_{\boldsymbol{k}\downarrow}(\boldsymbol{r})$ satisfy the relation:

$$u_{\boldsymbol{k}\downarrow}(\boldsymbol{r}+\boldsymbol{a}) = u_{\boldsymbol{k}\uparrow}(\boldsymbol{r}) . \tag{13.32}$$

This leads to the relation

$$\Psi_{\boldsymbol{k}\downarrow}(\boldsymbol{r}+\boldsymbol{a}) = \exp(i\boldsymbol{k}\boldsymbol{a})\Psi_{\boldsymbol{k}\uparrow}(\boldsymbol{r}) \tag{13.33}$$

From (6.3), (13.29) and (13.33) it is clear that $V_\uparrow^\gamma(\boldsymbol{k},\boldsymbol{k}')$ and $V_\downarrow^\gamma(\boldsymbol{k},\boldsymbol{k}')$ satisfy the following relation;

$$V_\uparrow^\gamma(\boldsymbol{k},\boldsymbol{k}') = \pm \exp(i\boldsymbol{K}\cdot\boldsymbol{a})V_\downarrow^\gamma(\boldsymbol{k},\boldsymbol{k}') , \tag{13.34}$$

where $\boldsymbol{K} = \boldsymbol{k} - \boldsymbol{k}' - \boldsymbol{q}$ and $\boldsymbol{a} = (a,0,0)$. The vector $\boldsymbol{K}$ takes a value of $m\boldsymbol{Q}_1 + n\boldsymbol{Q}_2 = (\pi/a, \pi/a, 0)m + (-\pi/a, \pi/a, 0)n$, with m and n being integers. And $\exp(i\boldsymbol{K}\cdot\boldsymbol{a})$ takes a value of $+1$ or -1, depending on whether a scattering process is normal or umklapp.

For the electron–phonon interaction matrix element in the case of an ordinary unit cell without the AF order, $\tilde{V}^\gamma(\boldsymbol{k},\boldsymbol{k}')$, we have

$$\tilde{V}_\uparrow^\gamma(\boldsymbol{k},\boldsymbol{k}')\tilde{V}_\downarrow^\gamma(-\boldsymbol{k},-\boldsymbol{k}') = \left|\tilde{V}_\uparrow^\gamma(\boldsymbol{k},\boldsymbol{k}')\right|^2 \tag{13.35}$$

and in this case the spectral function $\alpha^2 \tilde{F}_{\uparrow\downarrow}(\Omega,\theta,\theta')$ is always positive, i.e., attractive, for any combination of $\boldsymbol{k}$ and $\boldsymbol{k}'$. In the case of the AF unit cell which we are considering in this paper, however, we have

$$V_\uparrow^\gamma(\boldsymbol{k},\boldsymbol{k}')V_\downarrow^\gamma(-\boldsymbol{k},-\boldsymbol{k}') = \pm \exp(i\boldsymbol{K}\cdot\boldsymbol{a})\left|V_\uparrow^\gamma(\boldsymbol{k},\boldsymbol{k}')\right|^2 \tag{13.36}$$

and $\alpha^2 F_{\uparrow\downarrow}(\Omega,\theta,\theta')$ changes its sign according to the sign of $\pm\exp(i\boldsymbol{K}\cdot\boldsymbol{a})$.

Appendix F. D-wave Component of a Spectral Function and D-wave Superconductivity

In order to study the possibility of the occurrence of d-wave superconductivity, we have to solve the $\boldsymbol{k}$-dependent Eliashberg equation. Let a set of functions, $F_J(\boldsymbol{k})$'s, be complete and orthonormal when integrated on the Fermi surfaces. It is clear that $F_J(\boldsymbol{k})$'s reflect the symmetry of the band structure. In terms of this set of functions we can write

$$\Delta(\omega,\boldsymbol{k}) = \sum_J \Delta_J(\omega) F_J(\boldsymbol{k}) \tag{13.37}$$

$$Z(\omega,\boldsymbol{k}) = \sum_J Z_J(\omega) F_J(\boldsymbol{k}) \tag{13.38}$$

$$\alpha^2 F_{\uparrow\uparrow}(\Omega,\boldsymbol{k},\boldsymbol{k}') = \sum_{JJ'} \alpha^2 F_{\uparrow\uparrow JJ'}(\Omega) F_J(\boldsymbol{k}) F_{J'}(\boldsymbol{k}') \tag{13.39}$$

$$\alpha^2 F_{\uparrow\downarrow}(\Omega,\boldsymbol{k},\boldsymbol{k}') = \sum_{JJ'} \alpha^2 F_{\uparrow\downarrow JJ'}(\Omega) F_J(\boldsymbol{k}) F_{J'}(\boldsymbol{k}') \ . \tag{13.40}$$

With the use of these expansion coefficients of the gap function $\Delta(\omega,\boldsymbol{k})$, the renormalization function $Z(\omega,\boldsymbol{k})$ and the spectral functions $\alpha^2 F_{s,s'}(\Omega,\boldsymbol{k},\boldsymbol{k}')$, we obtain the following linearized Eliashberg equation for anisotropic superconductivity.

$$[1 - Z_J(\omega)]\omega = \sum_{J'} \int_{-\infty}^{\infty} d\omega' \int_0^{\infty} d\Omega \, \frac{\rho(\omega' Z(\omega'))}{\rho(E_F)} \alpha^2 F_{\uparrow\uparrow JJ'}(\Omega) I(\omega,\omega',\Omega) \tag{13.41}$$

$$[\Delta(\omega)Z(\omega)]_J = -\sum_{J'} \int_{-\infty}^{\infty} d\omega' \int_0^{\infty} d\Omega \, \frac{\rho(\omega' Z(\omega'))}{\rho(E_F)} \alpha^2 F_{\uparrow\downarrow JJ'}(\Omega) I(\omega,\omega',\Omega) \frac{\Delta_{J'}(\omega')}{\omega'} \tag{13.42}$$

where

$$I(\omega,\omega',\Omega) = \frac{1 - f(\omega')}{\omega - \Omega - \omega'} + \frac{f(\omega')}{\omega + \Omega - \omega'} \tag{13.43}$$

$$f(\omega) = \frac{1}{1 + \exp(\omega'/kT)} \tag{13.44}$$

Here $\rho(\omega)$ is the renormalized density of states of the hole carrier at energy ω. As we have already noted in Chap. 6, the spectral function which appears in the formula for the renormalization function (13.41), must be $\alpha^2 F_{\uparrow\uparrow JJ'}(\Omega)$ because this term contains the processes of virtual emissions and absorptions of various modes of phonons by a single electron, while the spectral function in (13.42) is $\alpha^2 F_{\uparrow\downarrow JJ'}(\Omega)$, which contains scattering processes of a pair of electrons from one pair state $(\boldsymbol{k}\uparrow, -\boldsymbol{k}\downarrow)$ to a different state $(\boldsymbol{k}'\uparrow, -\boldsymbol{k}'\downarrow)$.

From the two dimensional properties of LSCO, it seems to be an adequate approximation to take $\cos n\theta$'s as the complete and orthonormal set of functions, $F_J(\boldsymbol{k})$'s, where $\theta = \tan^{-1}(k_y/k_x)$. Then the linearized Eliashberg equation becomes,

$$[1 - Z_n(\omega)]\omega = \sum_{n'} \int_{-\infty}^{\infty} d\omega' \int_0^{\infty} d\Omega \, \frac{\rho(\omega' Z(\omega'))}{\rho(E_F)} \alpha^2 F_{\uparrow\uparrow nn'}(\Omega) I(\omega,\omega',\Omega) \tag{13.45}$$

$$[\Delta(\omega)Z(\omega)]_n = -\sum_{n'} \int_{-\infty}^{\infty} d\omega' \int_0^{\infty} d\Omega \, \frac{\rho(\omega' Z(\omega'))}{\rho(E_F)} \alpha^2 F_{\uparrow\downarrow nn'}(\Omega) I(\omega,\omega',\Omega) \frac{\Delta_{n'}(\omega')}{\omega'} \tag{13.46}$$

where

$$\Delta(\omega, \theta) = \sum_n C_n \Delta_n(\omega) \cos(n\theta) \quad (13.47)$$

$$Z(\omega, \theta, \theta') = \sum_n C_n Z_n(\omega) \cos(n\theta) \quad (13.48)$$

$$\alpha^2 F_{\uparrow\downarrow}(\Omega, \theta, \theta') = \sum_{nn'} C_n C_{n'} \alpha^2 F_{\uparrow\downarrow nn'}(\Omega) \cos n\theta \cos n\theta' . \quad (13.49)$$

where $C_n = 1/\sqrt{2\pi}$ for $n = 0$ and $C_n = 1/\sqrt{\pi}$ for $n \neq 0$. In Chap. 6 we have calculated the spectral function and shown that among the components of the spectral function, $\alpha^2 F_{\uparrow\uparrow nn'}(\Omega)$'s and $\alpha^2 F_{\uparrow\downarrow nn'}(\Omega)$'s, all terms are small and negligible except for $\alpha^2 F_{\uparrow\uparrow 0,0}(\Omega)$ and $\alpha^2 F_{\uparrow\downarrow 2,2}(\Omega)$. Following the expressions in Chap. 6, we include the normalization factor C_n in the expressions of $\alpha^2 F$ and for simplicity we use the notation $\alpha^2 F_{\uparrow\uparrow}^{(0)}(\Omega)$ and $\frac{1}{2}\alpha^2 F_{\uparrow\downarrow}^{(2)}(\Omega)$ for $\alpha^2 F_{\uparrow\uparrow 0,0}(\Omega)$ and $\alpha^2 F_{\uparrow\downarrow 2,2}(\Omega)$ respectively. Hereafter we use this notation. Then we obtain the following equation,

$$[1 - Z_0(\omega)]\omega = \int_{-\infty}^{\infty} d\omega' \int_0^{\infty} d\Omega \, \frac{\rho(\omega' Z(\omega'))}{\rho(E_F)} \alpha^2 F_{\uparrow\uparrow}^{(0)}(\Omega) I(\omega, \omega', \Omega) \quad (13.50)$$

$$\Delta_2(\omega) Z_0(\omega) = -\int_{-\infty}^{\infty} d\omega' \int_0^{\infty} d\Omega \, \frac{\rho(\omega' Z(\omega'))}{\rho(E_F)} \frac{\alpha^2 F_{\uparrow\downarrow}^{(2)}(\Omega)}{2} I(\omega, \omega', \Omega) \frac{\Delta_2(\omega')}{\omega'} \quad (13.51)$$

Note that the component of the spectral function which connects the s- and d-wave symmetry, $\alpha^2 F_{\uparrow\downarrow 2,0}(\Omega)$, vanishes from C_4 symmetry, and that the d-wave component of the spectral function $\alpha^2 F_{\uparrow\downarrow}^{(2)}(\Omega)$ is large while the s-wave component $\alpha^2 F_{\uparrow\downarrow}^{(0)}(\Omega)$ is negligibly small, as we have seen in Chap. 6. The d-wave component $\alpha^2 F_{\uparrow\downarrow}^{(2)}(\Omega)$ contributes to the formation of d-wave pairing as is known from (13.51). These results establish the appearance of the d-wave superconductivity in LSCO system.

14 Mechanism of High Temperature Superconductivity

In this chapter we explain how superconductivity occurs within the framework of the K–S model and clarify the key factors for determining the superconducting transition temperature T_c [28].

After an instructive discussion, we show that a characteristic electronic structure of the K–S model described in preceding chapters causes an anomalous effective electron–electron interaction between holes with different spins. One can see that the effective pairing interaction caused from the electron–phonon interaction on the K–S model becomes repulsive or attractive, in contrast to ordinary BCS cases in which the effective interaction is always *attractive*. In the following section we consider a simplified model derived from the electron–phonon interaction on the K–S model as an instructive method, and explain that d-wave symmetry is favored compared with s-wave symmetry in some parameter range for the K–S model.

We then discuss the effect of a finite spin-correlation length in the underlying AF-localized spin system. The finite size of the AF-correlation length observed in experiments suppresses the occurrence of superconductivity, and hence the interplay between the strong electron–phonon interaction and the local AF order in a finite size determines a superconducting transition temperature.

Because the present system has strong electron–phonon interaction, we have to go beyond the weak coupling calculation not to overestimate T_c. For this purpose, we employ a strong coupling treatment similar to McMillan's [197]. After the formalism is described briefly, we show the numerical results for hole concentration x-dependences of T_c and isotope effect α in the final section.

Since a considerable part of this chapter is devoted to the description of theoretical formulation, if readers have interests mainly in how the K–S model heads to d-wave pairing and in the calculated results of T_c and the isotope effect α in LSCO as a function of hole-concentration x, we suggest they read Sects. 14.1, 14.2 and 14.6.

14.1 Introduction

One might consider that the *electron–phonon interaction* can explain the high temperature superconductivity in the cuprates, like conventional superconducting systems. However a simple Cooper-pairing picture derived from the traditional point of view can hardly explain the d-wave symmetry observed in real cuprates since the Cooper pair interaction derived from the electron–phonon interaction always favors the formation of the s-wave symmetry. As seen in the calculation of the momentum-dependent spectral function $\alpha^2 F_{\uparrow\downarrow}(\Omega, \theta, \theta')$ in Chap. 13, however, in the K–S model the spatial difference of up- and down-spin wave functions due to the local AF order causes the d-wave pairing even for the electron–phonon interaction. In this chapter we will prove it rigorously.

Now let us describe why the traditional BCS pairing favors the s-wave pairing. In a usual metal, the effective electron–electron interaction derived from electron–phonon interaction is written as,

$$V(\boldsymbol{k}, \boldsymbol{\ell}) = g^*(\boldsymbol{k}, \boldsymbol{\ell}) g(\boldsymbol{k}, \boldsymbol{\ell}) \frac{2\omega_q}{(\varepsilon_{\boldsymbol{k}} - \varepsilon_{\boldsymbol{\ell}})^2 - \omega_q^2} , \qquad (14.1)$$

where, $\varepsilon_{\boldsymbol{k}}$ is the kinetic energy of a carrier hole with momentum $\boldsymbol{k}$. The process mentioned above corresponds to the scattering of electrons with momentum $\boldsymbol{k}$ and $-\boldsymbol{k}$ to momentum $\boldsymbol{\ell}$ and $-\boldsymbol{\ell}$ by the virtual emission and simultaneous absorption of a phonon with momentum $\boldsymbol{q} = \boldsymbol{\ell} - \boldsymbol{k}$, as illustrated in Fig. 14.1 (for the derivation of (14.1), see [60] for example). The factors $g(\boldsymbol{k}, \boldsymbol{\ell})$ and $g^*(\boldsymbol{k}, \boldsymbol{\ell})$ correspond to the emission and absorption of a phonon, respectively. Since the term $g^*(\boldsymbol{k}, \boldsymbol{\ell}) g(\boldsymbol{k}, \boldsymbol{\ell})$ is always positive, the effective interaction between a Cooper pair with relative momenta $2\boldsymbol{k}$ and $2\boldsymbol{\ell}$, with $\boldsymbol{k}$ and $\boldsymbol{\ell}$ both being near the Fermi surface, is always *attractive*. It is readily understood from the BCS equation that no sign change in the gap near the Fermi surface is favored. Thus the s-symmetry is realized.

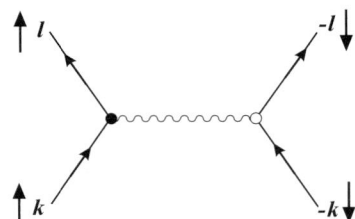

Fig. 14.1. Schematic picture of the effective electron–electron interaction derived from the electron–phonon interaction. The *black circle* represents the coupling constant between up-spin particles and phonons while the *white circle*, down-spin particles

14.2 Appearance of Repulsive Phonon-Exchange Interaction in the K–S Model

As described in the preceding section, the effective pairing interaction which is derived from electron–phonon interaction has a factor which takes the form of (*absorption process of a phonon*) × (*emission process of a phonon*). Because of the time-reversal symmetry of a system, the former term (*absorption process*) is generally the complex conjugate of the latter term (*emission process*).

From this fact it seems that the realization of d-wave pairing within the electron–phonon interaction scheme is hopeless. But if we examine closely how the coupling constant g of electron–phonon interaction was derived, we see that the explicit form of electron wave function plays an important role. Namely, if the spatial distributions of the wave function for up-spin carrier and down-spin carrier differ, there is no need for the relation (*absorption process*)* = (*emission process*) to be satisfied for scattering interactions of Cooper pairs. Indeed, for the electronic structure in the present K–S model, we shall find that the (*absorption process*) term is not simply the complex conjugate of the (*emission process*), but differs by a sign ± according to the momentum transfer q. Once the relation (*absorption process*) * = $-$ (*emission process*) is established for a particular momentum transfer q, then the effective interaction derived from this process gives rise to a repulsive term, which favors the sign change of the gap function during the scattering process caused by the virtual exchange of a particular momentum q phonon mentioned above. This is the scenario of the present theory. In the next subsection we derive the *selection rule* for what kind of exchange process yields the minus factor. Then in the following section, we will show that the interplay of such interaction and the Fermi surface structure of the K–S model really favors the d-wave symmetry.

14.2.1 The Selection Rule

In this subsection we show that the electron–phonon coupling constant depends on the spin of the carrier and the phonon momentum q within the framework of the K–S model. In general, the electron–phonon interaction comes from displacement of atoms from their equilibrium positions. It is well known that the scattering of a hole-carrier due to the displacement of atoms can generally be written as follows;

$$g(\bm{k},\bm{k}')_\sigma = \int \mathrm{d}\bm{r} v_{\bm{q}}(\bm{r}) \psi^*_{\bm{k}',\sigma}(\bm{r}) \psi_{\bm{k},\sigma}(\bm{r}) , \tag{14.2}$$

where $v_{\bm{q}}(\bm{r})$ denotes the scattering interaction term and $\psi_{\bm{k}^{(')},\sigma}(\bm{r})$ represents the normalized Bloch wavefunction with wavenumber $\bm{k}^{(')}$. Note that this is the most general expression for electron–phonon interaction. By Bloch's theorem, we can write $v_q(r) = e^{iq\cdot r}u_q(r)$, with $u_q(r)$ having the periodicity of a normal unit cell while $\psi_{k^{(\prime)},\sigma}(r) = e^{ik^{(\prime)}\cdot r}\phi_{k^{(\prime)},\sigma}(r)$, where $\phi_{k^{(\prime)}}(r)$ has the periodicity of an *AF unit cell*. Then using the AF periodicity, we can rewrite the integral (14.2) as follows.

$$g(k,k')_\sigma = \sum_{R_n}\int_\Delta dr e^{iq\cdot(r-R_n)}u_q(r-R_n)\psi^*_{k',\sigma}(r-R_n)\psi_{k,\sigma}(r-R_n)$$
$$= \sum_{R_n}e^{-i(k-k'+q)\cdot R_n}\int_\Delta dr e^{i(k-k'+q)\cdot r}u_q(r)\phi^*_{k',\sigma}(r)\phi_{k,\sigma}(r) \ .$$
(14.3)

Here R_n's run through all the AF-unit-vectors and Δ denotes the integral over the AF-unit cell. The sum $\sum_{R_n} e^{i(k-k'+q)\cdot R_n}$ yields the term $\sum_Q \delta_{k-k'+q\ Q}$, which means the conservation of pseudo-momentum with the AF-periodicity during scattering processes, where Q's represent the reciprocal vectors of the AF-periodicity. Then (14.3) becomes

$$g(k,k')_\sigma = \sum_Q \delta_{k-k'+q\ Q}\int_\Delta dr e^{i(k-k'+q)\cdot r}u_q(r)\phi^*_{k',\sigma}(r)\phi_{k,\sigma}(r) \ . \quad (14.4)$$

Now we show that the electron–phonon coupling constants $g(k,k')_\uparrow$ and $g(k,k')_\downarrow$ are spin-dependent based on the K–S model. First, let us establish the relation between $\phi_{k,\downarrow}(r)$ and $\phi_{k,\uparrow}(r)$.

In the preceding chapter we obtained the following equality (see (13.33) in Appendix E, Chap. 13) for the Bloch functions in the K–S model:

$$\psi_{k^{(\prime)},-\sigma}(r) = e^{ik\cdot u_1}\psi_{k^{(\prime)},\sigma}(r-u_1) \ . \quad (14.5)$$

Hence we have
$$\phi_{k^{(\prime)},-\sigma}(r) = \phi_{k^{(\prime)},\sigma}(r-u_1) \ . \quad (14.6)$$

Noticing that $u_q(r)$ has the periodicity of a normal unit cell, the integral for the $g(k,k')_{-\sigma}$ is now written as,

$$g(k,k')_{-\sigma} = \sum_Q \delta_{k-k'+q\ Q}\int_\Delta dr e^{i(k-k'+q)\cdot r}u_q(r)\phi^*_{k',-\sigma}(r)\phi_{k,-\sigma}(r)$$
$$= \sum_Q \delta_{k-k'+q\ Q}$$
$$\times \int_\Delta dr e^{i(k-k'+q)\cdot r}u_q(r)\phi^*_{k',\sigma}(r-u_1)\phi_{k,\sigma}(r-u_1)$$
$$= \sum_Q \delta_{k-k'+q\ Q}e^{i(k-k'+q)\cdot u_1}$$
$$\times \int_{\Delta+u_1} dr e^{i(k-k'+q)\cdot r}u_q(r)\phi^*_{k',\sigma}(r)\phi_{k,\sigma}(r)$$
$$= e^{i(k-k'+q)\cdot u_1}g(k,k')_\sigma \ . \quad (14.7)$$

Here, on going from the second row to the third row in (14.7), we have changed an integral variable from $\boldsymbol{r}$ to $\boldsymbol{r}+\boldsymbol{u_1}$ and have used the normal unit cell periodicity for the function $u_{\boldsymbol{q}}(\boldsymbol{r})$, where $\boldsymbol{u}_1$ is a vector connecting with Cu–O–Cu distance. On going from the third row to the last row in (14.7), we have used the fact that integrals of periodic functions over the periodicity are unique and do not depend on the choice of a region of periodicity.

From the pseudo-momentum conservation we obtain $\boldsymbol{k}-\boldsymbol{k}'+\boldsymbol{q}=\boldsymbol{K}$, with $\boldsymbol{K}$ being a reciprocal vector in the AF Brillouin zone. That is, $\boldsymbol{K}=(n\pi/a, m\pi/a)$, where n, m are integers which satisfy the condition $n+m$ being even. Since $\boldsymbol{u}_1=(a,0)$, one can derive the following relation from (14.7);

$$g(\boldsymbol{k},\boldsymbol{k}')_{-\sigma} = (-1)^n g(\boldsymbol{k},\boldsymbol{k}')_\sigma \ . \tag{14.8}$$

Note that both n and m take even numbers or odd numbers at the same time, so that (14.8) preserves the tetragonal symmetry. On the derivation of (14.8), we have only used (14.6). Then one may wonder if the same conclusion is drawn for systems with the normal periodicity by formally folding a system to the AF-periodicity. Equation (14.8) itself surely holds for the system with the normal periodicity. But from the pseudo-momentum conservation law, we have

$$\boldsymbol{k} - \boldsymbol{k}' + \boldsymbol{q} = \boldsymbol{K} \ . \tag{14.9}$$

Thus no peculiar thing occurs in the case of the normal periodicity, pseudo-momentum conservation requires that $\boldsymbol{K}$'s appearing in (14.9) must be the reciprocal vectors of the *normal* periodicity. Hence the factor $(-1)^n$ in (14.8) is always equal to unity. On the other hand, (14.8) yields the non-trivial result of having opposite sign between $g(\boldsymbol{k},\boldsymbol{k}')_\uparrow$ and $g(\boldsymbol{k},\boldsymbol{k}')_\downarrow$ for some $\boldsymbol{k}$, $\boldsymbol{k}'$ and $\boldsymbol{q}$ in any systems with AF-periodicity and the different spatial distributions between up- and down-spin electrons, such as spin density wave (SDW) states. But SDW states are of course insulating so that we do not have much interest in the context of the superconductivity.

Based on the K–S model, we have shown that the electron-coupling constants of different spin carriers can really differ by a sign, but there is still an ambiguity for the expression of (14.8). The ambiguity comes from the fact that $\boldsymbol{k}$'s have the AF-periodicity, while $\boldsymbol{q}$'s have the normal periodicity. This ambiguity is taken away by requiring that $\boldsymbol{k}$'s are confined to the first AF-Brillouin zone (AF-BZ), while $\boldsymbol{q}$'s are confined to the first Brillouin zone of the normal periodicity (normal BZ). For the scattering from momentum $\boldsymbol{\ell}$-state to $\boldsymbol{k}$-state, we have two kinds of process for the same phonon branch, reflecting the fact that phonons have the normal unit cell periodicity while carriers have the AF-unit cell periodicity.

In the two processes the pseudo-momenta $\boldsymbol{k}-\boldsymbol{\ell}+\boldsymbol{q}$ differ from the zero vector by the reciprocal lattice vector $\pm\boldsymbol{Q}_1$ with $\boldsymbol{Q}_1=(\pi/a,\pi/a)$, $\pm\boldsymbol{Q}_2$ with $\boldsymbol{Q}_2=(-\pi/a,\pi/a)$ or $\pm\boldsymbol{Q}_1\pm\boldsymbol{Q}_2$ (or, equal to the zero vector). Then we have sign change of the electron–phonon interaction $g_\downarrow(\boldsymbol{\ell},\boldsymbol{k})$ and $g_\uparrow(\boldsymbol{\ell},\boldsymbol{k})$ which occurs for the case where the difference of pseudo-momentum equals to $\pm\boldsymbol{Q}_1$

166   14 Mechanism of High Temperature Superconductivity

or $\pm\boldsymbol{Q}_2$. Because of the pillar-type structures of Fermi surfaces on the K–S electronic model obtained in Chap. 11 (see Fig. 11.3, Fig. 11.4), in most of scattering processes which are important for the formation of Cooper pairs, the sign change occurs.

As seen from the figure, two processes corresponding to the same pseudo-momentum transfer $\boldsymbol{k}-\boldsymbol{\ell}$ in the AF-periodicity always have different signs for dominant interactions in Cooper pairing. For this sign change we can derive a more simple rule from Fig. 14.2. Let us say that a wave vector belongs to A-Brillouin zone when it can be placed in the first AF Brillouin zone by the translation of AF-reciprocal lattice vector $\boldsymbol{K} = n\boldsymbol{Q}_1 + m\boldsymbol{Q}_2$ with $n+m$ being an even number while it belongs to B-Brillouin zone when $n+m$ is an odd number. Then we see from the figure that for each momentum transfer $\boldsymbol{k}-\boldsymbol{\ell}$, the momenta of phonons involved in the two processes may be placed in the A- or the B-Brillouin zone. If we call the process with phonon momentum in the A-Brillouin zone a "Normal "-process N and the other "Umklapp"-process U, we can say that the above two contributions create the sign change in the electron–phonon interaction, such as $g(\boldsymbol{k},\boldsymbol{\ell})_{\sigma\mathrm{N}} - g(\boldsymbol{k},\boldsymbol{\ell})_{\sigma\mathrm{U}}$ or $g(\boldsymbol{k},\boldsymbol{\ell})_{\sigma\mathrm{U}} - g(\boldsymbol{k},\boldsymbol{\ell})_{\sigma\mathrm{N}}$, which varies when $\boldsymbol{\ell}$ changes with $\boldsymbol{k}$ being fixed.

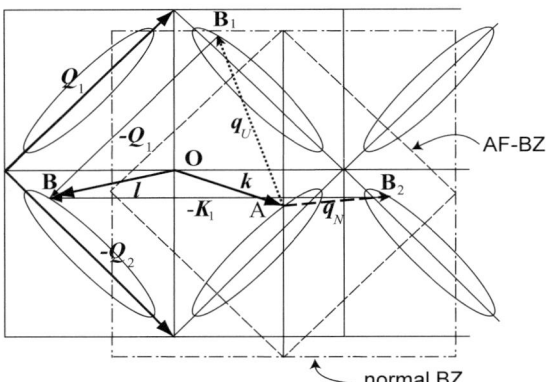

Fig. 14.2. Schematic picture for the selection rule, where a scattering process from the point A to B is chosen as an example. Note that there are two phonon processes (*dashed* and *dotted arrows*) within the same phonon branch. Under the antiferromagnetic periodicity, B_1 and B_2 points are equivalent to point B. Vectors $\boldsymbol{Q}_1$ and $\boldsymbol{Q}_1 - \boldsymbol{Q}_2$ in the figure correspond to $\boldsymbol{K}$ vector in the text (see 14.9). As a consequence, the scattering process represented by the *dashed arrow* gives an attractive interaction while the *dotted arrow* yields a repulsive interaction in this case. Thus there exist two contributions to the scattering A to B, which have different sign. Competition between two processes A→B_1→B and A→B_2→B determines the sign of attractive or repulsive interaction. Note that if the momentum of phonon $\boldsymbol{q}_\mathrm{N}$ is in the first AF-Brillouin zone, then the other $\boldsymbol{q}_\mathrm{U}$ is not

Then we readily see from the figure that the effective interaction has a strong ℓ dependence, leading to the sign change of the effective interaction.

14.2.2 Occurrence of the d-wave Symmetry

In this subsection we describe how the $d_{x^2-y^2}$-symmetry pairing observed in experiments [56, 59, 72, 75] occurs in the present model within the framework of the weak coupling, i.e., within the second order perturbation theory. More realistic strong coupling treatment will be given in the following section, but it is instructive to understand the mechanism intuitively in the present section. For readers who have interests in the calculated results of superconducting properties in real cuprates and in the comparison between theoretical and experimental results, they may jump to Sect. 14.6, by skipping from this subsection to Sect. 14.5.

In the present model, we show that the occurrence of the $d_{x^2-y^2}$-symmetry pairing depends on two factors. One is the appearance of effective repulsive interactions due to phonon-exchange as described in the preceding subsection and the other is the characteristic shape of the Fermi-surface derived by Kamimura and Ushio [112, 113], which consists of four sections of small, elongated ellipsoidal form as illustrated in Fig. 11.3 in Chap. 11. We also note that the van Hove singularity point is placed at the G_1 point $(\pi/a, 0)$, where the bent parts of the Fermi surface have a large value of the partial density of the states. As a result, scattering from these bent regions of the Fermi surface to other bent regions, is a main contribution to the formation of a Cooper pair. Then for the $d_{x^2-y^2}$-wave pairing, the sign change of the effective interaction $V(\boldsymbol{k}, \boldsymbol{\ell})$ is favored depending on placements of the two vectors such as shown in Fig. 14.3.

To see this, let us consider the following weak coupling-equation for the $\boldsymbol{k}$-dependent gap function $\Delta(\boldsymbol{k}, T)$ at temperature T.

$$\Delta(\boldsymbol{k}, T) = -\sum_{\boldsymbol{\ell}} V(\boldsymbol{k}, \boldsymbol{\ell}) \frac{\Delta(\boldsymbol{\ell}, T)}{2E(\boldsymbol{\ell})} \tanh[\beta E(\boldsymbol{\ell})/2], \qquad (14.10)$$

$$E(\boldsymbol{k}) = \sqrt{\xi(\boldsymbol{k})^2 + \Delta(\boldsymbol{k}, T)^2}, \ \beta = 1/T.$$

Here, $\xi(\boldsymbol{k})$ is the kinetic energy of a carrier hole with momentum $\boldsymbol{k}$ measured from the chemical potential μ and $V(\boldsymbol{k}, \boldsymbol{\ell})$ represents the effective interaction. From Fig. 14.3 and (14.10) we easily see that if $V(\boldsymbol{k}, \boldsymbol{\ell})$ in Fig. 14.3 is attractive and $V(\boldsymbol{k}', \boldsymbol{\ell})$ is repulsive, $d_{x^2-y^2}$-gap is really favored.

Now let us investigate what kind of symmetry is favored by various types of effective interactions V's by simplified weak coupling model calculations. The $V(\boldsymbol{k}, \boldsymbol{\ell})$ term and $\Delta(\boldsymbol{\ell})$ term are of course continuous functions of $\boldsymbol{\ell}$, but for simplicity we smear out its momentum dependence by taking the sum over the regions around the Fermi surface as shown in Fig. 14.4, i.e., we set

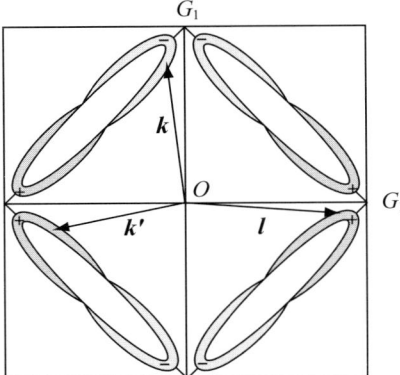

Fig. 14.3. Schematic picture of scattering from ℓ to k' by the pairing interaction. These two processes mainly contribute to the formation of the superconducting gap. If the scattering from ℓ to k is attractive and the one ℓ to k' is repulsive, the formation of the $d_{x^2-y^2}$-gap is favored together with the sign change of $d_{x^2-y^2}$-gap symmetry. Amplitude and sign of the $d_{x^2-y^2}$-gap function is shown schematically

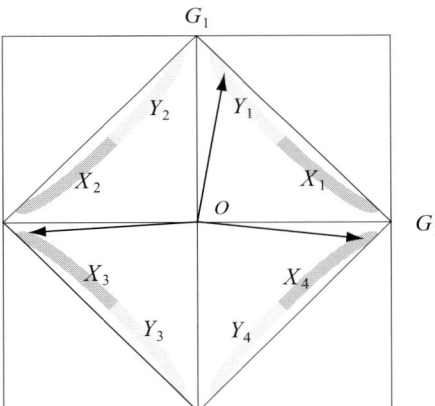

Fig. 14.4. Intuitive illustration of the averaging of the pairing interaction described in the text. For $d_{x^2-y^2}$-symmetry, the gap function is positive in the dark-coloured regions and negative in the light-coloured regions. Pair interaction is effective in these thin ellipsoidal regions

14.2 Appearance of Repulsive Phonon-Exchange Interaction

$$\Delta_x = \int_{X_1} \mathrm{d}\boldsymbol{k}\Delta(\boldsymbol{k},T)/S ,$$

$$\Delta_y = \int_{Y_1} \mathrm{d}\boldsymbol{k}\Delta(\boldsymbol{k},T)/S ,$$

$$V_1 = \sum_{i=1}^{4} \int_{X_i} \mathrm{d}\boldsymbol{\ell} \int_{X_1} \mathrm{d}\boldsymbol{k} V(\boldsymbol{k},\boldsymbol{\ell})/S^2 ,$$

$$V_2 = \sum_{i=1}^{4} \int_{Y_i} \mathrm{d}\boldsymbol{\ell} \int_{X_1} \mathrm{d}\boldsymbol{k} V(\boldsymbol{k},\boldsymbol{\ell})/S^2, \text{ where, } S \text{ is given as}$$

$$S = \int_{X_i} \mathrm{d}\boldsymbol{\ell} = \int_{Y_i} \mathrm{d}\boldsymbol{\ell} ,$$

where the integral regions X_{1-4} and Y_{1-4} are shown in the figure. The regions X_{1-4} correspond to the parts of thin shells around the Fermi surface of thickness $2\omega_\mathrm{D}$ (ω_D being the Debye frequency) defined by $|\xi(\boldsymbol{k})| < \omega_\mathrm{D}$ where the gap function takes a positive value, while Y_i's correspond to the part where the gap is negative in the case of $\mathrm{d}x^2 - y^2$-symmetry. This means that we reduce the problem to a problem of solving the averaged value of gap amplitude by averaging the interactions. From the tetragonal symmetry of the system we have,

$$\Delta_x = \pm \Delta_y . \tag{14.11}$$

Here, the plus sign corresponds to s-symmetry, while the minus corresponds to $\mathrm{d}x^2 - y^2$-symmetry. Simply writing the absolute values of Δ_x and Δ_y as Δ, now the gap equation is reduced to

$$1 = -\rho_\mathrm{F}(V_1 \pm V_2) \int_0^{\omega_\mathrm{D}} \mathrm{d}\xi \frac{\tanh[\beta E/2]}{E(\xi)} , \tag{14.12}$$

$$E(\xi) = \sqrt{\xi^2 + \Delta(T)^2} ,$$

where ω_D is the Debye frequency and ρ_F is the density of the states at the Fermi level. By introducing non-dimensional quantities

$$\lambda_1 = -\rho_\mathrm{F} V_1, \ \lambda_2 = -\rho_\mathrm{F} V_2, \ \lambda_\pm = \lambda_1 \pm \lambda_2 , \tag{14.13}$$

Equation (14.12) becomes

$$1 = \lambda_\pm \int_0^{\omega_\mathrm{D}} \mathrm{d}\xi \frac{\tanh[\beta E(\xi)/2]}{E(\xi)} , \tag{14.14}$$

which coincides with the form of the ordinary BCS equation. Thus we can immediately obtain the transition temperature T_c as follows

$$T_\mathrm{c}^{(\pm)} = 1.13 \omega_\mathrm{D} e^{-1/\lambda_\pm} , \tag{14.15}$$

where the sign $\pm$ corresponds to s- or d-symmetry. Here we should stress that the gap (14.14) has a solution only when $\lambda_\pm > 0$. This means that the overall interaction must be attractive. When both λ_+ and λ_- take positive values, the criterion of whether s- or d-gap occurs is given as follows. Namely, if $T_c^{(+)} > T_c^{(-)}$ is satisfied, we have the s-symmetry; if $T_c^{(+)} < T_c^{(-)}$ is satisfied, we have the d-symmetry.

As we have noted, the largest contributions for the formation of the gap function comes from regions nearby the G_1 point at which the band dispersion takes the van-Hove singularity. Then the averaged interactions $V_{1,2}$'s are roughly estimated from values of $V(\boldsymbol{k}^{(\prime)}, \boldsymbol{\ell})$'s in Fig. 14.3. If the value $V(\boldsymbol{k}', \boldsymbol{\ell})s$ in Fig. 14.3 is negative, i.e., the interaction is *attractive*, we have $V_1 < 0$, while if $V(\boldsymbol{k}, \boldsymbol{\ell})$'s is positive, i.e., *repulsive* interaction, we have $V_2 > 0$. Then we have

$$\lambda_1 > 0, \ \lambda_2 < 0, \quad (14.16)$$

and as a result

$$\lambda_- > \lambda_+, \ \lambda_- = \lambda_1 + |\lambda_2| > 0, \quad (14.17)$$

so that

$$T_c^{(-)} > T_c^{(+)}. \quad (14.18)$$

This shows stability of d-symmetry compared with s-symmetry in the above mentioned parameter region. In the present simplified model, clearly the $\lambda_2 = 0$ line corresponds to the boundary separating the regions where s-symmetry or d-symmetry is preferred.

By directly solving the gap (14.10) numerically, we obtain essentially the same result. The result of numerical calculations for the ground state is shown in Fig. 14.5. In the numerical calculation we adopt a simplified form for the effective interaction. That is,

$$V(\boldsymbol{k}, \boldsymbol{\ell}) = -\Lambda_\mathrm{N}, \ \text{or} \ -\Lambda_\mathrm{U}. \quad (14.19)$$

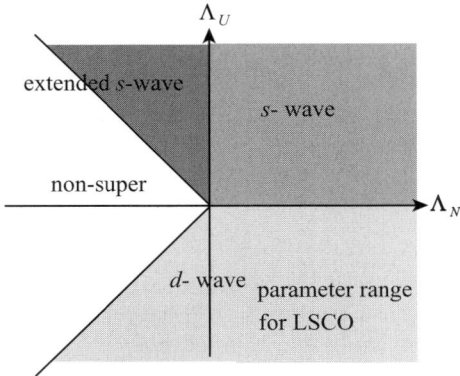

Fig. 14.5. Numerical result for simplified pair interaction with coupling constants Λ_N and Λ_U. Here, positive values of Λ's correspond to attractive interactions

Here, the selection of Λ_N or Λ_U depends on the momentum transfer $\boldsymbol{q}=\boldsymbol{k}-\boldsymbol{\ell}$. Subscript N represents normal scattering and U represents Umklapp scattering in the AF Brillouin zone. Signs of Λ's are given to coincide with the conventional notation, i.e., a positive value of Λ corresponds to the attractive interaction. From Fig. 14.5 we readily see that the result of the much simplified treatment of (14.14) is reproduced. One exception is the occurrence of strongly anisotropic s-wave solution (extended s-wave), which has almost 0-gap amplitude around Δ points $(\pm\pi/a, \pm/a)$ [28].

In this way the problem of whether we can explain $d_{x^2-y^2}$-symmetry within the present model is reduced to calculating Λ's from realistic electron–phonon interaction model. The calculation has been done by Ushio and Kamimura using a semi-empirical method developed by Motizuki et al. [198], and they found that λ_1 and λ_2 almost have the same magnitude with opposite sign, where λ_2 is negative [112]. Thus we conclude that in the present model, superconductivity with $d_{x^2-y^2}$-gap symmetry appears. As is well known, the effective pairing interaction derived from the electron–phonon interaction in conventional metallic systems is always attractive, *the sign change* of the effective interaction really reflects the unique feature of electronic states of the K–S model.

As we have already mentioned, in the numerical calculation we used a much simplified form for the effective interaction instead of the form derived from the second order perturbation shown in (14.1). We did not use this interaction in the present calculation because it is known that such a treatment overestimates the Cooper pair coupling formation. A proper treatment should include effects which tend to suppress superconductivity, and in general we have to take account of the electron–phonon interaction up to the infinite order of the perturbation expansion.

14.3 Suppression of Superconductivity by Finiteness of the Anti-Ferromagnetic Correlation Length

So far we have investigated whether superconductivity occurs by the electron–phonon interaction based on the K–S model. In the preceding section we assumed the existence of static anti-ferromagnetic order, but in a realistic system such a long-range static order is not observed, reflecting the low-dimensionality of copper-oxide systems. Instead, the spin-correlation length λ_s is finite, and thus we have to consider the finite-size effect of a metallic state due to the finite anti-ferromagnetic spin correlation length. This causes a finite lifetime effect on a hole-carrier excitation. In addition to a finite lifetime effect due to the inelastic interaction between quasi-particle excitations, a quasi-particle excitation has a finite lifetime due to the dynamical 2D-AF fluctuation in the localized spin system, as we described in detail in Fig. 11.2 in Sect. 11.2.

14.3.1 Influence of the Lack of the Static, Long-Ranged AF-Order on the Electronic Structure

As mentioned in Sect. 8.2.1, it is known from neutron scattering experiments that two-dimensional anti-ferromagnetic correlation exists throughout the carrier concentration region where the superconducting transition occurs [159, 171, 199]. The result of neutron inelastic scattering experiments indicates that each CuO_2-layer consists of regions of average size $\lambda_s \times \lambda_s$ within which the AF-order due to the localized spins exists. Boundaries of AF-ordered regions are not of classical form and they fluctuate in the sense of quantum mechanics. The scale of AF-ordered region λ_s can be estimated from neutron inelastic scattering experiments as an inverse of the half width of the anti-ferromagnetic incommensurate peaks, which corresponds to the AF-correlation length [37, 53, 142]. Because of the finiteness of AF-ordered range, we can no longer consider that Bloch functions discussed in the preceding chapter are well defined over a whole CuO_2-layer. However, we can still draw a quasi-particle excitation picture by recognizing that the hole-carrier excitation has a considerably long lifetime due to the fluctuation of the AF-background even in low temperatures (see Fig. 11.2, Fig. 14.6). Namely, the quasi-particle-excitation is well defined in one AF-ordered region in which a hole-carrier can move freely. In other words, it has a much longer "mean free path" than the spin-correlation length, i.e., $\ell_0 > \lambda_s$. Then letting $v = \|\boldsymbol{v_p}\|$, $\boldsymbol{v_p} = \nabla_{\boldsymbol{p}} \xi(\boldsymbol{k})$ be the velocity of the hole-carrier, the lifetime τ_h of the excitation is determined from the relation,

$$v\tau_h = \ell_0 . \tag{14.20}$$

In calculating T_c of cuprates based on the K–S model, we have to take this finite lifetime effect into account.

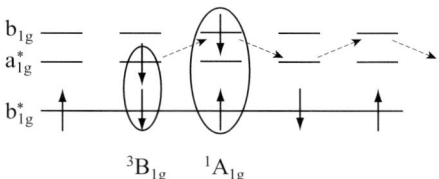

Fig. 14.6. Schematic diagram of AF-fluctuation effect. A quasi-particle excitation is well defined in the region of size $\ell_0 \times \ell_0$

14.3.2 Suppression of Superconductivity Due to the Finite Lifetime Effect

Let us recall the form of the effective interaction derived from electron–phonon interaction in (14.1). This term comes from a exchange process of a virtual phonon, and it has retardation effect, i.e., the interaction is not instantaneous. And without any disturbance, this virtual process is effective to infinite time range. Namely, the interaction is attractive when the condition $|\epsilon(\boldsymbol{k}) - \epsilon(\boldsymbol{k}')| < \omega_{\boldsymbol{q}}$ is satisfied and from the energy-time uncertainty this means that the interaction which occurs in the time range $T > \tau_{\mathrm{D}} = 1/\omega_{\mathrm{D}}$ is always attractive. But in the present model, because of the finite lifetime effect mentioned in the preceding subsection, we no longer have effective pairing interaction of infinite time range. That is, interaction of a time range longer than $\tau_{\mathrm{hF}} \equiv \ell_0/v_{\mathrm{F}}$ between two quasi-particles with the Fermi velocity v_{F}, is ineffective because of the finite correlation length of the localized AF-spin system, causes changes of the electronic structure of hole-carrier states. Hence two hole-carriers are not able to couple beyond the time τ_{hF}.

Let us apply this finite lifetime effect. We require any virtual phonon exchange process in the present model must be completed within the time range $\Delta t = \tau_{\mathrm{hF}}$. Then the gap (14.14) is rewritten as

$$1 = \lambda_- \int_{\omega_{\mathrm{hF}}}^{\omega_{\mathrm{D}}} \mathrm{d}\xi \frac{\tanh[\beta E(\xi)/2]}{E(\xi)}, \qquad (14.21)$$

where $\omega_{\mathrm{hF}} = 1/\tau_{\mathrm{hF}}$ represents the lower cut-off parameter which comes from the finite lifetime effect, and for the T_c-equation we have

$$1 = \lambda_- \int_{\omega_{\mathrm{hF}}}^{\omega_{\mathrm{D}}} \mathrm{d}\xi \frac{\tanh[\beta_c \xi/2]}{\xi}, \quad \text{where } \beta_c = 1/T_c. \qquad (14.22)$$

The velocity of a hole-carrier varies with its momentum, and the cut-off constant ω_{hF} depends on the momentum of the hole-carrier in general. But what we are studying now is the simplified (14.22), so we will treat ω_{hF} as a parameter and ignore its momentum dependence. In the numerical calculation we have treated the velocity of hole-carriers to be a constant Fermi-velocity v_{F}, which is the averaged value of velocities over the Fermi surface for the hole concentration $x = 0.15$, and we treated the mean free path of hole-carriers ℓ_0 discussed in the preceding section as a parameter. Then ω_{hF}, which is given by $\omega_{\mathrm{hF}} = v_{\mathrm{F}}/\ell_0$, was treated as a parameter in the calculation for Fig. 14.7.

The result of numerical calculations of (14.22) is given in Fig. 14.7. Here the electron–phonon coupling constant λ_- is also treated as a parameter. It is seen from the figure that the rate of decrease of the transition temperature T_c with decreasing ℓ_0 is slow when ℓ_0 is large enough. Then when ℓ_0 decreases to a certain value (depending on the coupling constant λ_-) T_c begins to drop rapidly. The value of ω_{hF} at which T_c vanishes is determined from the following equation.

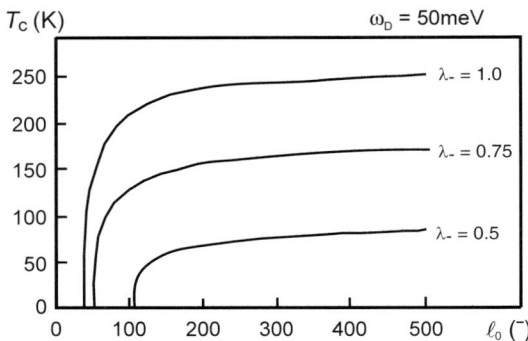

Fig. 14.7. Calculated result of T_c for three values of electron–phonon coupling constant λ_- for simplified interaction with the finite AF-correlation length effect. ℓ_0 gives the lower cut-off parameter $\omega_{\rm hF}$ by $\omega_{\rm hF} = v_{\rm F}/\ell_0$ (see the text)

$$1 = \lambda_- \int_{\omega_{\rm hF}}^{\omega_{\rm D}} d\xi \frac{1}{\xi} = \ln\left[\frac{\omega_{\rm D}}{\omega_{\rm hF}}\right] . \qquad (14.23)$$

From this, we obtain

$$\omega_{\rm hF} = \omega_{\rm D} e^{-1/\lambda_-} . \qquad (14.24)$$

The result is rather a trivial one. If the pairing coupling constant is small, then T_c vanishes for a small $\omega_{\rm hF}$, while in the $\lambda_- \to \infty$ limit, T_c does not vanish unless $\omega_{\rm hF}$ reaches $\omega_{\rm D}$ [27].

14.4 Strong Coupling Treatment of Conventional Superconducting System

In this section we describe briefly the Green's function method which is used for conventional strong coupling systems (for detailed treatment, see [200, 201]). As is well known, this method enables us to calculate various physical quantities without calculating eigenfunctions of the system: We can compute physical quantities including the effect of interaction Hamiltonian up to the infinite order of perturbation, without dropping the most important parts of perturbation series. First, we discuss the Green's function method in the normal state of a usual metallic system. Then we show how the method is applied to a conventional superconducting system.

14.4.1 Green's Function Method in the Normal State

In the Heisenberg picture, the one-body Green's function of a carrier system is defined as

14.4 Strong Coupling Treatment of Conventional Superconducting System

$$G(\boldsymbol{x}, \boldsymbol{x}', t) = \begin{cases} -i \ll \psi_\sigma(\boldsymbol{x}, t)\psi_\sigma^\dagger(\boldsymbol{x}', 0) \gg & \text{if } t > 0 \\ i \ll \psi_\sigma^\dagger(\boldsymbol{x}', 0)\psi_\sigma(\boldsymbol{x}, t) \gg & \text{if } t \leq 0 \end{cases} \quad (14.25)$$

where, $\psi_\sigma(\boldsymbol{x}, t)$ is the field operator of the spin σ carrier system and $\ll \cdots \gg$ denotes the thermal average, namely, $\ll A \gg = \text{Tr}\{Ae^{-\beta(H-\mu N)}\}/\text{Tr}\{e^{-\beta(H-\mu N)}\}$. Here, we used the grand-canonical ensemble of the system for convenience. We also introduce the non-perturbed Green's function G_0 by

$$G_0(\boldsymbol{x}, \boldsymbol{x}', t) = \begin{cases} -i \ll \psi_{H_0}(\boldsymbol{x}, t)\psi_{H_0}^\dagger(\boldsymbol{x}', 0) \gg & \text{if } t > 0 \\ i \ll \psi_{H_0}^\dagger(\boldsymbol{x}', 0)\psi_{H_0}(\boldsymbol{x}, t) \gg & \text{if } t \leq 0 \end{cases} \quad (14.26)$$

where H_0 is the Hamiltonian for the free carrier system and $\psi_{H_0}(\boldsymbol{x}, t)$ means that the field operator is evolving by the non-perturbed H_0. We dropped spin indices just for simplicity. Then the equation for G (the Feynman–Dyson equation) in the normal state is given as

$$\begin{aligned} G(\boldsymbol{x}, \boldsymbol{x}', t) &= G_0(\boldsymbol{x}, \boldsymbol{x}', t) \\ &+ \int\int\int\int G(\boldsymbol{x}, \boldsymbol{x_2}, t - t_2) dt_2 d\boldsymbol{x_2} \\ &\quad \times \Sigma(\boldsymbol{x_1}, \boldsymbol{x_2}, t_2 - t_1) dt_1 d\boldsymbol{x_1} G_0(\boldsymbol{x_1}, \boldsymbol{x}', t_1) , \end{aligned} \quad (14.27)$$

where the Σ-term represents the so-called irreducible self-energy part. Once the Σ-term is known, it is quite easy to determine G. To understand this, let us express the above equation in the $(\omega, \boldsymbol{k})$ space, i.e., the Fourier transformed space. Then the equation is written as,

$$G(\omega, \boldsymbol{k}) = G_0(\omega, \boldsymbol{k}) + G_0(\omega, \boldsymbol{k})\Sigma(\omega, \boldsymbol{k})G(\omega, \boldsymbol{k}) \quad (14.28)$$

By dividing both sides of the equation by $G(\omega, \boldsymbol{k})G_0(\omega, \boldsymbol{k})$, the equation reduces to

$$\frac{1}{G(\omega, \boldsymbol{k})} = \frac{1}{G_0(\omega, \boldsymbol{k})} - \Sigma(\omega, \boldsymbol{k}) , \quad (14.29)$$

so that computing the $\Sigma(\omega, \boldsymbol{k})$ term becomes a main task in this formalism. The irreducible self energy part $\Sigma(\boldsymbol{k}, \omega)$ is formally written as perturbation expansion by the interaction Hamiltonian. Every term in this expansion series has one-to-one correspondence to graphical representation, well known as *Feynman graphs*, and with the aid of Feynman graph expansion, we can discuss the Feynman–Dyson equation more intuitively (see Fig. 14.8). The wavy line contained in each graph represents the phonon Green's function $D(\boldsymbol{x}, \boldsymbol{x}', t)$ which is defined as

$$D_\alpha(\boldsymbol{x}, \boldsymbol{x}', t) = \begin{cases} -i \ll \varphi_{H_0}(\boldsymbol{x}, t)\varphi_{H_0}(\boldsymbol{x}', 0) \gg & \text{if } t > 0 \\ i \ll \varphi_{H_0}(\boldsymbol{x}', 0)\varphi_{H_0}(\boldsymbol{x}, t) \gg & \text{if } t \leq 0 \end{cases} \quad (14.30)$$

where $\varphi_\alpha(\boldsymbol{x}, t)$ is the phonon's field operator in Heisenberg picture and α denotes the phonon branch. We have to construct the Feynman equation for

176   14 Mechanism of High Temperature Superconductivity

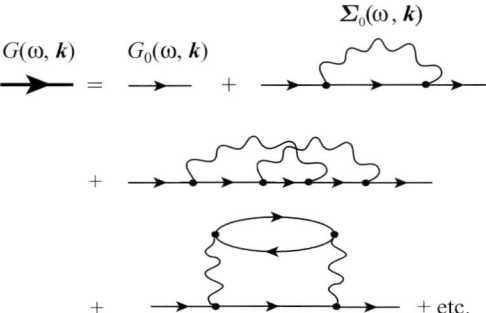

Fig. 14.8. The Feynman diagram for the carrier system. The thick line represents the complete Green's function and the thin line denotes the non-perturbed Green's function, while the wavy line represents the phonon Green's function

(a)

(b)

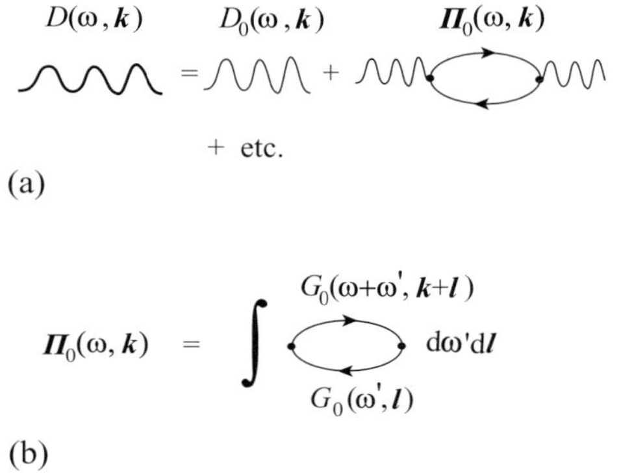

Fig. 14.9. (a) The Feynman diagram for the phonon system. (b) The bubble diagram

phonons as well as that of hole carriers. It is known that in the Feynman equation for phonons, the processes expressed in Fig. 14.9(a) have the most important contributions to the perturbation expansions. So we just omit other diagrams. Then the Feynman equation in the Fourier transformed space is reduced to

$$\frac{1}{D(\omega, \boldsymbol{k})} = \frac{1}{D_0(\omega, \boldsymbol{k})} - g^2 \Pi_0(\omega, \boldsymbol{k}) , \qquad (14.31)$$

where g is the electron–phonon coupling constant and $D_0(\omega, \boldsymbol{q})$ is the non-perturbed phonon Green's function, and $\Pi_0(\omega, \boldsymbol{k})$, the so called bubble diagram expressed in Fig. 14.9(b), is defined as

$$\Pi_0(\omega, \boldsymbol{k}) = -2i \int \frac{d\boldsymbol{q}}{(2\pi)^D} \int \frac{d\omega'}{2\pi} G(\omega', \boldsymbol{q}) G(\omega + \omega', \boldsymbol{k} + \boldsymbol{q}) . \qquad (14.32)$$

14.4 Strong Coupling Treatment of Conventional Superconducting System

Here, D denotes the dimension of the system. Thus in the case of high-temperature superconductivity, we set $D = 2$.

It is known that the $\Pi_0(\omega, \boldsymbol{k})$-term contributes to a shift of phonon frequency and introduces finite lifetime for a phonon excitation. The former effect can dramatically change the phonon's real Green's function from the non-perturbed one, which appears in the case of charge density wave (CDW) instability, known as the Kohn anomaly. On the other hand, the latter effect is generally harmless in the present treatment. Since experimental results show that there is no CDW instability except for the very special value of hole concentration ratio $x = 0.125$, for simplicity we omit not only finite lifetime effects but also treat phonon dispersions as *renormalized* in the following numerical calculation, i.e., the effects of shift of phonon frequencies are already included in the expression of dispersion relations. We note that this method is widely used in past theoretical studies for strong-coupled superconductors [197, 201]. Then within the present approximation, the perturbed phonon Green's functions have the same form as the non-perturbed ones.

Let us return to the carrier system. Graphically, the irreducible self-energy term Σ is expressed by all the graphs, each of which represents consequent virtual emission and absorption of phonons. But such graphs that are unconnected or can be separated into two parts by cutting the graph at a point on a line which represents carrier's Green's function must be omitted. These properties are the reason we call the term *irreducible* (see Fig. 14.10). Introducing just one more term called the irreducible vertex part Γ, we can then rewrite the irreducible self energy term in terms of the carrier and phonon Green's functions and the irreducible vertex part as follows.

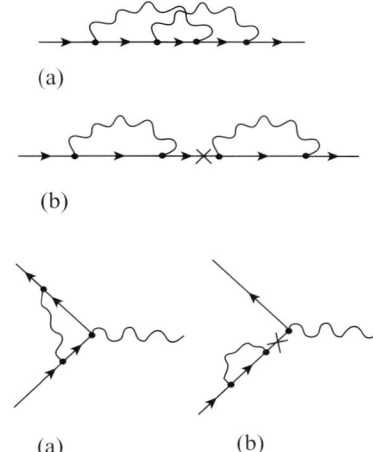

Fig. 14.10. (a) Graphical explanation of the irreducible self energy part and the irreducible vertex part and (b) reducible parts. If we cut the graphs of (b) at points marked by crosses, these graphs become unconnected

$$\Sigma(\omega, \boldsymbol{k}) = \sum_\alpha \int \frac{\mathrm{d}\boldsymbol{q}}{(2\pi)^D} \int \frac{\mathrm{d}\omega'}{2\pi} g^*_{\alpha,\boldsymbol{q}} D_\alpha(\omega', \boldsymbol{q}) G(\omega - \omega', \boldsymbol{k} - \boldsymbol{q}) \Gamma(\omega', \boldsymbol{q}) ,$$
(14.33)

where $g^*_{\alpha,\boldsymbol{q}}$ is the coupling constant of electron–phonon interaction. As seen from Fig. 14.10(b), the irreducible vertex part is regarded as the completely dressed interaction term. The above expression means that the irreducible self-energy term $\Sigma(\omega, \boldsymbol{k})$, which includes all the effects of the electron–phonon interaction, is given by the same form as that of the second order perturbation theory, if we use the complete Green's function G instead of the bare Green's function G_0 and use the completely dressed interaction Γ instead of the bare interaction $g_{\alpha,\boldsymbol{q}}$ appearing in the formula of the second order perturbation theory.

Now we can proceed to the next step. In almost any case we cannot solve the Feynman equation rigorously, so we must make some approximations to the equation. As for electron–phonon coupled system, there is a crude but efficient approximation first pointed out by Migdal [202]. That is, use the *bare* electron–phonon interaction $g_{\alpha,\boldsymbol{q}}$ instead of the vertex part $\Gamma(\omega', \boldsymbol{q})$ in the equation for the self energy part just described above. This approximation is justified under certain kinds of conditions: 1. Phonon momentum $\boldsymbol{q}$ should not be small. 2. The Fermi surface should not be small. Clearly, the contribution of phonon modes with small momentum $\boldsymbol{q}$ *is* small compared with all the contributions, so that we do not have to worry about the first condition. But the second condition seems really annoying if we want to apply Migdal's approximation to the K–S model, since the Fermi surface shown in Chap. 11 is indeed *small* in usual sense. Ushio and Kamimura have derived this small surface by assuming the existence of static AF-structure, but as we mentioned in this chapter, in a real system the AF-structure is fluctuating dynamically and it has only finite correlation length. This fact means that, though the Fermi surfaces in the K–S model are small geometrically, it is large in the sense of electron–phonon interaction because for the phonon scattering processes, the wave vectors of hole-carriers must be confined in the first Brillouin zone, as we have seen in Chap. 11. Thus Migdal's treatment is valid for the present case.

Adopting Migdal's approximation, the Feynman equation is now reduced to the following equation

$$\Sigma(\omega, \boldsymbol{k}) = \mathrm{sum}_\alpha \int \frac{\mathrm{d}\boldsymbol{q}}{(2\pi)^D} \int \frac{\mathrm{d}\omega'}{2\pi} g^*_{\alpha,\boldsymbol{q}} D_\alpha(\omega', \boldsymbol{q}) g_{\alpha,\boldsymbol{q}} G(\omega - \omega', \boldsymbol{k} - \boldsymbol{q}) . \quad (14.34)$$

Note that the Green's function in the right hand side of the equation contains the self-energy part implicitly through (14.29), so that the above equation should be solved self-consistently. This fact means that the equation includes an infinite order of perturbations, if not all of them.

14.4.2 Application of the Green's Function Method to a Superconducting State

So far we have discussed the Green's function method in the normal state of a usual metallic system. Now we apply the Green's function method to the superconducting state first derived by Eliashberg [203]. First, we introduce the *anomalous* Green's function $F(x, x', t)$ [204] by

$$F(x, x', t) = \begin{cases} -i \ll \psi_\uparrow(x, t)\psi_\downarrow(x', 0) \gg & \text{if } t > 0 \\ i \ll \psi_\downarrow(x', 0)\psi_\uparrow(x, t) \gg & \text{if } t \leq 0 . \end{cases} \quad (14.35)$$

Then the equation for the superconducting state is written in terms of G_σ and F in a compact form as first derived by Nambu [205]. Since the Nambu notation is purely a mathematical one, in the present chapter we describe the method just briefly and we leave the detailed discussion to the references already mentioned ([200, 201]). In the Fourier transformed-space, the Green's function for momentum k frequency ω and spin $\uparrow$ and its complex conjugate for momentum $-k$ frequency ω and spin $\downarrow$ together with the anomalous Green's function (and its complex conjugate) form a single 2×2 matrix-formed Green's function G as follows.

$$\boldsymbol{G}(\omega, \boldsymbol{k}) = \begin{pmatrix} G(\omega, \boldsymbol{k}) & F(\omega, \boldsymbol{k}) \\ F^*(\omega, \boldsymbol{k}) & G(\omega, \boldsymbol{k}) \end{pmatrix} . \quad (14.36)$$

Then, introducing the matrix-formed self-energy part $\boldsymbol{\Sigma}(\omega, \boldsymbol{k})$, the following relation holds as in the case of normal state (c.f. (14.29)),

$$\boldsymbol{G}^{-1}(\omega, \boldsymbol{k}) = \boldsymbol{G}_0^{-1}(\omega, \boldsymbol{k}) - \boldsymbol{\Sigma}(\omega, \boldsymbol{k}) , \quad (14.37)$$

where A^{-1} denotes the inverse of matrix A, and $\boldsymbol{G}_0(\omega, \boldsymbol{k})$ represents the non-interacting Green's function, i.e.,

$$\boldsymbol{G}_0(\omega, \boldsymbol{k}) = \begin{pmatrix} G_0(\omega, \boldsymbol{k}) & 0 \\ 0 & G_0^*(\omega, \boldsymbol{k}) \end{pmatrix} . \quad (14.38)$$

It is well known that any 2×2 matrix is always written as a linear superposition of the unit matrix $\tau_0 = I$ and the Pauli matrices τ_i, $i = 1-3$. Explicit forms of τ matrices are given by

$$\tau_0 = \begin{pmatrix} 1 & 0 \\ 0 & 1 \end{pmatrix}, \quad \tau_1 = \begin{pmatrix} 0 & 1 \\ 1 & 0 \end{pmatrix}, \quad \tau_2 = \begin{pmatrix} 0 & -i \\ i & 0 \end{pmatrix}, \quad \tau_3 = \begin{pmatrix} 1 & 0 \\ 0 & -1 \end{pmatrix} . \quad (14.39)$$

Using this notation, the matrix-formed self-energy term $\boldsymbol{\Sigma}(\omega, \boldsymbol{k})$ appearing in (14.37) can generally be written as

$$\boldsymbol{\Sigma}(\omega, \boldsymbol{k}) = \omega(1 - Z(\omega, \boldsymbol{k}))\tau_0 + \chi(\omega, \boldsymbol{k})\tau_3 + \phi_1(\omega, \boldsymbol{k})\tau_1 + \phi_2(\omega, \boldsymbol{k})\tau_2 , \quad (14.40)$$

where τ_0 denotes the unit matrix and functions Z, χ, ϕ_1 and ϕ_2 can be arbitrary functions of $\boldsymbol{k}$ and ω at this stage. From this point of view, solving the Feynman–Dyson equation is the same as determining the form of four functions Z, χ, ϕ_1 and ϕ_2. The form of the coefficient of the unit matrix τ_0 given in the above equation is for later convenience; we could have just written the coefficient of τ_0 such as $X(\omega, \bm{k})$. Since $X(\omega, \bm{k})$ is known to be the odd function of ω from general argument, $Z(\omega, \bm{k})$ is the even function of ω, where $Z \equiv 1$ corresponds to the non-perturbed case. Note that $\bm{\Sigma} \equiv 0$ in the non-perturbed case.

The factor Z in (14.40) is called the *renormalization factor*, and it is known to lower the transition temperature, and it does not appear in the second-order perturbation theory. This is the main reason we have adopted the Green's function method. Without the renormalization factor Z, there arises a risk of overestimating T_c, which we *must* avoid. Other functions χ, ϕ_1 and ϕ_2 have physical significance too. Roughly speaking, χ corresponds to the Hatree–Fock term in the second order perturbation theory, and in the later calculations we neglect it since we can include this effect in one-electron energy dispersion and it does not affect the superconductivity of the system. ϕ_1 and ϕ_2 correspond to the gap function of the superconductor. More precisely the function $\Delta(\omega, \bm{k}) = \sqrt{(\phi_1(\omega, \bm{k})/Z(\omega, \bm{k}))^2 + (\phi_2(\omega, \bm{k})/Z(\omega, \bm{k}))^2}$ corresponds to the frequency-dependent gap function of the system (for example, see [200]). And we also note here that the gauge invariance of the present system enables us to set $\phi_2 \equiv 0$. Namely, the total Hamiltonian of the system does not change under the gauge transformation $T(\theta)$,

$$H = T(\theta) H T(\theta)^\dagger, \qquad (14.41)$$

where $T(\theta) = \exp[i\theta\tau_3]$ is the (global) gauge transformation. Then if $\bm{\Sigma}(\omega, \bm{k})$ gives the irreducible self energy of the system, $T(\theta)\bm{\Sigma}(\omega, \bm{k})T(\theta)^\dagger$ can also be the solution of the Feynman–Dyson equation and we can eliminate the τ_2-term in $\bm{\Sigma}(\omega, \bm{k})$ from any solution by choosing the appropriate value of θ. So hereafter let us consider the self energy term which has no τ_2-term. Then the self-energy part is determined from the Feynman equation as in the case of normal state, and it is given as follows.

$$\bm{\Sigma}(\omega, \bm{k}) = \sum_\alpha \int \frac{d\bm{q}}{(2\pi)^D} \int \frac{d\omega'}{2\pi} g(\alpha, \bm{q})^* \tau_3 D_\alpha(\omega', \bm{q}) \bm{G}(\omega - \omega', \bm{k} - \bm{q}) g(\alpha, \bm{q}) \tau_3. \qquad (14.42)$$

Here, D_α is the phonon Green's function for the phonon mode α as before, and the Migdal's approximation has already been made in deriving the (14.42). That is, in the present treatment the irreducible vertex part is simply substituted by the bare vertex term $g(\alpha, \bm{q})\tau_3$ instead of the complete one. The Green's function in the right hand side of (14.42) depends on the self-energy part $\bm{\Sigma}$ through (14.37), and hence basically, we are able to solve the equation for $\bm{\Sigma}$ in (14.42) self-consistently, once the phonon Green's functions D_α and the electron–phonon coupling constants $g_{\alpha, \bm{q}}$ are given.

14.4.3 Inclusion of Coulomb Repulsion

So far we have not discussed the effect of Coulomb repulsion between hole-carriers, which of course must be included in the calculation of T_c. The bare Coulomb interaction is instantaneous and long-ranged, but after we take into account the screening effect it is no more instantaneous nor long-ranged. Inclusion of Coulomb interaction has a long history for conventional superconducting metals (see [201], for example). Most of them are based on the free electron model, with sufficiently large electron density, either of which of course does not hold in the present case, where the electronic structure has a highly tight binding character with low carrier density. But because of this fact we can get rid of the somewhat complicated arguments made in references mentioned above. Using the tight-binding structure from the beginning means that much of the Coulomb interaction is already included in the determination of many-body-effect including band structures in the K–S model. That is, in the K–S model hole-carriers with up and down spins occupy different orbitals of a_{1g}^* and b_{1g} symmetry inside the same CuO_6 octahedron. Further, owing to the low density of hole-carriers, two hole-carriers with opposite spins are well separated. In fact, the results of the exact-diagonalization study of the K–S model described in Chap. 9 has shown that the calculated radial distribution function reveal the highest probability when two holes are separated by 8.7 Å, which is close to the experimental results on the coherent length of a Cooper pair. Thus we conclude that the Coulomb interaction is small enough in the electronic states of the K–S model, and hence we can reasonably neglect the Coulomb repulsion parameter μ_c in the present calculation.

14.4.4 T_c-Equation in the Strong Coupling Model

An application of the Green's function method in conventional superconductors was first discussed in detail by McMillan [197]. In his treatment he reduced the Feynman equation (14.42) with four variables ω and $\boldsymbol{k}$ to the equation with one variable ω. We will follow his argument but there is one important difference in the present treatment. In other words, we have to treat d-wave symmetry instead of s-wave symmetry, which McMillan had considered. It appears that the extension of McMillan's treatment to the (14.42) for arbitrary temperature is difficult but applicable to the T_c-equation, as we shall discuss in the following. Thus we concentrate on the application of McMillan's method to the T_c-equation. In the following, we derive the general form of a T_c-determining equation first.

If the temperature T included implicitly in (14.42) is close enough to the superconducting transition temperature T_c, then the amplitude of anomalous Green's function $F(\omega, \boldsymbol{k})$ is small. In the limit $T \to T_c$, the equation is reduced to the linear equation with respect to the anomalous Green's function, which corresponds to the off-diagonal part of the matrix-formed Green's function

G. The linearized equation is then given as follows. We first write the form of the self-energy part Σ

$$\Sigma(\omega, \boldsymbol{k}) = \omega(1 - Z(\omega, \boldsymbol{k}))\tau_0 - \chi(\omega, \boldsymbol{k})\tau_3 - \delta\phi(\omega, \boldsymbol{k})\tau_1 , \qquad (14.43)$$

where $\delta\phi$ is infinitesimal quantity around $T \sim T_c$ and since χ and Z functions are continuous function of temperature T, they can be set to those of $T_c T_c$. From (14.37), we have

$$\boldsymbol{G}(\omega, \boldsymbol{k}) = (\boldsymbol{G}_0^{-1}(\omega, \boldsymbol{k}) - \boldsymbol{\Sigma}(\omega, \boldsymbol{k}))^{-1} . \qquad (14.44)$$

Recalling the form of the non-perturbed Green's function $\boldsymbol{G}_0$ in (14.38) and the from of the self-energy part $\boldsymbol{\Sigma}$ in (14.43), *the linearized* Green's function $\boldsymbol{G}^l(\omega, \boldsymbol{k})$ which have $\delta\phi$ function up to the first order is easily obtained as

$$\boldsymbol{G}^l(\omega, \boldsymbol{k}) = \frac{\omega Z(\omega, \boldsymbol{k})\tau_0 + (\xi_{\boldsymbol{k}} + \chi(\omega, \boldsymbol{k}))\tau_3 + \delta\phi(\omega, \boldsymbol{k})\tau_1}{\omega^2 Z^2(\omega, \boldsymbol{k}) - (\xi_{\boldsymbol{k}} + \chi(\omega, \boldsymbol{k}))^2 + i\delta} , \qquad (14.45)$$

where δ denotes a positive infinitesimal. Then the linearized Feynman–Dyson equation becomes

$$\boldsymbol{\Sigma}(\omega, \boldsymbol{k}) = \int \frac{d\boldsymbol{q}}{(2\pi)^D} \int \frac{d\omega'}{2\pi} \frac{1}{\rho_F} \int d\Omega \alpha^2 F(\Omega, \boldsymbol{k}, \boldsymbol{\ell}) D(\omega - \omega', \Omega) \tau_3 \boldsymbol{G}^l(\omega', \boldsymbol{\ell}) \tau_3 . \qquad (14.46)$$

Here ρ_F denotes the density of states of the carrier system at the Fermi level, and the $\alpha^2 F$-term in the above equation comes from all the contributions of the phonon modes, and is defined as

$$\alpha^2 F(\Omega, \boldsymbol{k}, \boldsymbol{\ell}) = \rho_F \sum_\alpha \delta(\Omega - \omega_\alpha(\boldsymbol{k} - \boldsymbol{\ell})) g_\alpha^*(\boldsymbol{k}, \boldsymbol{\ell}) g_\alpha(\boldsymbol{k}, \boldsymbol{\ell}) , \qquad (14.47)$$

where ρ_F denotes the density of the states at the Fermi level. As we have seen, the electron–phonon coupling constant g depends on the spin of the hole-carrier in the K–S model. At present, we neglect that fact for simplicity. After the final form of the T_c equation is obtained, we give the prescription for the spin-dependent case. If one wishes to treat the problem properly, an interaction term proportional to the unit matrix τ_0 also appears. The coefficient corresponds to the case $g_\downarrow = -g_\uparrow$, and the proper treatment leads us to precisely the same form of T_c-equation we give in the following.

Now let us proceed. The function $D(\omega, \Omega)$ in (14.46) has the same form as the Green's function of phonon with frequency Ω. It is easy to see that the delta function in the definition of the $\alpha^2 F$-term reproduces (14.42), when $\alpha^2 F(\Omega, \boldsymbol{k}, \boldsymbol{\ell})$ is integrated over Ω.

Following McMillan, we change the integral variables for the $\boldsymbol{\ell}$ integration. First, we consider the $\chi(\omega, \boldsymbol{\ell})$ part. It is known that in general this term does not have remarkable ω-dependence. Therefore, comparing the coefficient of the τ_3-term for the bare Green's function with that of the complete Green's function, we understand the effect of χ-term as a deviation of the carrier-energy-dispersion from the bare one. (Change of $\xi(\bm{k})$ to $\xi(\bm{k}) + \chi(*, \bm{k})$.) We include this effect in the dispersion relation from the beginning; we need not consider the χ-term anymore. Hereafter we denote $\xi(\bm{k})$ the renormalized carrier hole energy measured from the chemical potential. Then the volume element $d\bm{\ell}/(2\pi)^D$ can be written as $\rho(\xi)d\xi dS(\bm{\ell})$, where $\rho(\xi)$ denotes the density of the states over an equal-energy surface defined by $\xi(\bm{\ell}) = \xi$, and $dS(\bm{\ell})$ is an area element on this surface. As a result (14.46) is now expressed by changing of variables described above as follows.

$$\omega(1 - Z(\omega, \xi, \bm{k}))\tau_0 - \delta\phi(\omega, \xi, \bm{k})\tau_1$$
$$= \int \frac{d\omega'}{2\pi} \int d\xi' \frac{\rho(\xi')}{\rho_F} \int dS(\bm{\ell}) \int d\Omega \alpha^2 F(\Omega, \xi, \bm{k}, \xi', \bm{\ell})$$
$$\times D(\omega - \omega', \Omega)\tau_3 \frac{\omega' Z(\omega', \xi', \bm{\ell})\tau_0 + \delta\phi(\omega, \xi', \bm{\ell})\tau_1}{\omega'^2 Z^2(\omega', \xi', \bm{\ell}) - \xi'^2 + i\delta}\tau_3 \ . \quad (14.48)$$

For the case of an isotropic s-wave which McMillan had treated, the ξ'-integration in (14.48) can be made analytically with several assumptions. In the present K–S model we have d-wave symmetry which is of course highly anisotropic and we cannot ignore $\bm{\ell}$-dependence appearing in the right-hand side of (14.48). Hence we have to modify somewhat the original McMillan's treatment. We shall discuss this in the following section.

14.5 Application of McMillan's Method to the K–S Model

In order to apply McMillan's method to highly anisotropic cases, following Allen [201, 206], we introduce a set of basis functions $\{f_n(\xi, \bm{k}), n = 1, 2, \cdots, \xi(\bm{k}) = \xi\}$ that are orthogonal and complete on the equal energy surface $\xi(\bm{k}) = \xi$ in the sense

$$\int_{\xi(\bm{k})=\xi} f_n^*(\xi, \bm{k}) f_m(\xi, \bm{k}) dS(\bm{k}) = \delta_{nm} \ , \quad (14.49)$$

$$\sum_n f_n(\xi, \bm{k})^* f_n(\xi, \bm{\ell}) = \delta_\xi(\bm{k} - \bm{\ell}) \ . \quad (14.50)$$

Here the subscript attached to the delta function means that it is the delta function for the equal energy surface. Allen [206] introduced these functions in analogy with spherical harmonics $Y_{lm}(\bm{\ell})/(4\pi)^{1/2}$, and called them "Fermi surface harmonics". But we note here that there is no need to require that functions $f_n(\xi, \bm{\ell})$, $n = 1, 2, \cdots$ satisfy some special kind of condition coming from the geometric structure of equal energy surfaces, e.g., to be the eigenfunctions of the Laplace-Beltrami operator on an equal energy surfaces $\xi(\bm{\ell}) = \xi$. We just need the ortho-normality of the set of functions

$\{f_n(\xi, \boldsymbol{\ell}), \ n = 1, 2, \cdots\}$. Now the terms appearing in (14.46) can be expanded by f_n's using the orthonormal relation (14.49) and (14.50). Namely,

$$\boldsymbol{\Sigma}(\omega, \xi(\boldsymbol{\ell})) = \sum_n \boldsymbol{\Sigma}_n(\omega, \xi) f_n(\xi, \boldsymbol{\ell}) \tag{14.51}$$

$$Z(\omega, \xi(\boldsymbol{\ell})) = \sum_n Z_n(\omega, \xi) f_n(\xi, \boldsymbol{\ell}) \tag{14.52}$$

$$\delta\phi(\omega, \xi(\boldsymbol{\ell})) = \sum_n \delta\phi_n(\omega, \xi) f_n(\xi, \boldsymbol{\ell}) \tag{14.53}$$

$$\frac{1}{\omega^2 Z^2(\omega, \xi(\boldsymbol{\ell})) - \xi_\ell^2 + i\delta} = \sum_n \left(\frac{1}{\omega^2 Z^2(\omega, \xi) - \xi^2 + i\delta}\right)_n f_n(\xi, \boldsymbol{\ell}) \tag{14.54}$$

$$\alpha^2 F(\Omega, \boldsymbol{\ell}, \boldsymbol{q}') = \sum_n \alpha^2 F_{n,m}(\Omega, \xi, \xi') f_n^*(\xi, \boldsymbol{\ell}) f_m(\xi', \boldsymbol{q}') \tag{14.55}$$

Now (14.46) is written as follows.

$$\begin{aligned}\boldsymbol{\Sigma}_n(\omega, \xi) &= \sum_m \int \frac{\mathrm{d}\omega'}{2\pi} \int \mathrm{d}\xi' \frac{\rho(\xi')}{\rho_{\mathrm{F}}} \\ &\quad \times \int \mathrm{d}\Omega \alpha^2 F_{n,m}(\Omega, \xi, \xi') D(\omega - \omega', \Omega) \tau_3 \boldsymbol{G}_m^l(\omega', \xi') \tau_3 \ ,\end{aligned} \tag{14.56}$$

$$\boldsymbol{\Sigma}_n(\omega, \xi) = \omega(\delta_{0n} - Z_n(\omega, \xi))\tau_0 - \delta\phi_n(\omega, \xi)\tau_1 \ , \tag{14.57}$$

$$\begin{aligned}\boldsymbol{G}_m^l(\omega', \xi') &= \sum_{jk} \mathcal{D}(\xi')_{mjk}\{\omega' Z_j(\omega', \xi)\tau_0 - \delta\phi_j(\omega', \xi')\tau_1\}_j \\ &\quad \times \left(\frac{1}{\omega'^2 Z(\omega', \xi')^2 - \xi'^2 + i\delta}\right)_k .\end{aligned} \tag{14.58}$$

Here, $\mathcal{D}_{mjk}$'s are the constants which are defined as

$$f_j(\xi, \boldsymbol{\ell}) f_k(\xi, \boldsymbol{\ell}) = \sum_m \mathcal{D}_{mjk} f_m(\xi, \boldsymbol{\ell}) \ , \tag{14.59}$$

and we took f_0 to be constant on each equal energy surface, i.e., totally isotropic function with respect to the shape of equal energy surface $\xi(\boldsymbol{\ell}) = \xi$.

Equation (14.58) is now linearized in terms of the pair function $\delta\phi$, but it still has complicated form which clearly comes from f_n-expansion. If the renormalization factor $Z(\omega, \xi(\boldsymbol{\ell}))$ is totally isotropic, i.e., if it does not depend on points on the surface $\xi = \xi(\boldsymbol{\ell})$ then the equation is reduced to a simple form as we shall see in the following and we adopt this assumption. As we have mentioned, we can take any set of functions $\{f_n\}$ as long as f_n's satisfy (14.49) and (14.50). The main assumption here is that the Z-term is totally isotropic in terms of the equal-energy surface. Then, as we mentioned, if we take $f_0 = \kappa$ to be constant on the same surface we have

14.5 Application of McMillan's Method to the K–S Model

$$Z(\omega, \xi(\boldsymbol{\ell})) = \kappa Z_0(\omega, \xi) , \qquad (14.60)$$

with f_0 *being constant on* $\xi = \xi(\boldsymbol{\ell})$. Then the function $1/(\omega^2 Z(\omega, \xi(\boldsymbol{\ell}))^2 - \xi(\boldsymbol{\ell})^2 + i\delta)$ appearing in (14.54) is also totally isotropic so that it is also proportional to f_0, i.e.,

$$\frac{1}{\omega^2 Z^2(\omega, \xi(\boldsymbol{\ell})) - \xi(\boldsymbol{\ell})^2 + i\delta} = \kappa \left(\frac{1}{(\omega^2 Z(\omega, \xi))^2 - \xi^2 + i\delta} \right)_0 . \qquad (14.61)$$

Then (14.58) becomes

$$\boldsymbol{G}_m^l(\omega', \xi') = \sum_j \mathcal{D}(\xi')_{mj0} \{ \omega' \frac{Z(\omega', \xi')}{\kappa} \delta_{j,0} \tau_0 - \delta\phi_j(\omega', \xi') \tau_1 \}$$

$$\times \frac{1}{\kappa} \frac{1}{\omega'^2 Z^2(\omega', \xi') - \xi'^2 + i\delta} . \qquad (14.62)$$

By the definition of $\mathcal{D}(\xi)_{mjk}$ in (14.59) we readily see

$$\mathcal{D}_{mj0} = \kappa \delta_{mj} . \qquad (14.63)$$

Now (14.62) becomes

$$\boldsymbol{G}_m^l(\omega', \xi') = \{ \omega' \frac{Z(\omega', \xi')}{\kappa} \delta_{m0} \tau_0 - \delta\phi_m(\omega', \xi') \tau_1 \} \frac{1}{\omega'^2 Z^2(\omega', \xi') - \xi'^2 + i\delta} . \qquad (14.64)$$

Inserting the above expression for $\boldsymbol{G}_m^l$ appearing in (14.56) we finally have the equation,

$$\boldsymbol{\Sigma}_n(\omega, \xi) = \sum_m \int \frac{d\omega'}{2\pi} \int d\xi' \frac{\rho(\xi')}{\rho_F} \int d\Omega \alpha^2 F_{n,m}(\Omega, \xi, \xi') D(\omega - \omega', \Omega)$$

$$\times \{ \omega' Z(\omega', \xi') \delta_{m0}) I - \delta\phi_m(\omega', \xi') \tau_1 \}$$

$$\times \frac{1}{\omega'^2 Z(\omega', \xi')^2 - \xi'^2 + i\delta}, \qquad (14.65)$$

$$\boldsymbol{\Sigma}_n = \omega (1 - Z(\omega, \xi)) \delta_{n0} I - \delta\phi_n(\omega, \xi) \tau_1 , \qquad (14.66)$$

which corresponds to anisotropic version of McMillan's equation. Since it is known that the ω-dependence of the Z-term, particularly its value at $\omega = 0$, has the most important effect on suppression of superconductivity [197], the above mentioned approximation to ignore the $\boldsymbol{k}$-dependence of the Z-term makes sense and hereafter we will concentrate on solving (14.65).

Now the form of (14.65) allows us to follow McMillan's method, that is to perform the ξ'-integral first and reduce the problem of solving a one variable ω-dependent equation. On performing the ξ'-integral in (14.65), we use the fact that, in general, the $\alpha^2 F$-term in (14.65) does not have significant ξ, ξ'-dependence so that we can set the $\alpha^2 F$-term to be constant as regards to variables ξ and ξ'. This means that the left hand side of (14.65) has no longer ξ-dependence. That is, both the Z-term and $\delta\phi$ are no longer ξ-dependent. We also ignore the ω-dependence of the Z-term. It is an even function of ω and has its largest value at $\omega = 0$, and in the limit $\omega \to \infty$ it approaches unity (for example, see [207]). So, substituting $Z(\omega)$ by $Z(0)$ nearby $\omega \sim 0$ does not affect the result much. When $\omega \gg 0$, other terms appearing in the right hand side of (14.65) become very small so that even in the region $\omega \gg 0$ where $Z(\omega)$ and $Z(0)$ should differ, the *substitution* $Z(0) \to Z(\omega)$ *to all ω-interval* is not expected to change the quantitative results of (14.65). (For similar kinds of treatments, see [208], for example.)

Then the left hand side of (14.65) now depends only on ω and the ξ'-integral of the right hand side can be done analytically just the same as McMillan's treatment and given as

$$\Sigma_n = \sum_m \int d\omega' \frac{\rho(Z(0)\omega')}{\rho_F} \int d\Omega \alpha^2 F_{n,m}(\Omega) I(\omega, \Omega, \omega') \frac{\omega' Z(0)\tau_0 - \delta\phi_m(\omega')\tau_1}{|\omega' Z(0)|} . \tag{14.67}$$

Here the function $I(\omega, \Omega, \omega')$ is defined by

$$I(\omega, \Omega, \omega') = \frac{N(\Omega) + 1 - f(\omega')}{\omega + i\delta - \Omega - \omega'} + \frac{N(\Omega) + f(\omega')}{\omega + i\delta + \Omega - \omega'} , \tag{14.68}$$

where $N(\Omega)$ denotes Bose distribution function $N(\Omega) = 1/(e^{\beta_c \Omega} - 1)$ and $f(\omega')$ denotes the Fermi distribution function $f(\omega') = 1/(1 + e^{\beta_c \omega'})$. Since we are now considering the T_c equation, we set $\beta_c = 1/T_c$ in the above expression. The derivation of (14.67) from (14.65) by ξ'-integral is a purely mathematical one, and it is the same for conventional strong-coupling treatments. (For details, see [201], Sect. 12. As seen from the reference, we need *one more* mathematical step for the derivation of (14.67) from (14.65), namely the spectral representation of the $\boldsymbol{G}^l$-term. Then the origin of the temperature-dependent term I in (14.67) becomes clear.)

Now we divide the gap equation into two parts. One for the determination of the renormalized factor $Z(0)$ and the other for the determination of the infinitesimal gap function $\delta\phi_n$. Namely,

$$(1 - Z(0))\omega = \sum_m \int d\omega' \frac{\rho(Z(0)\omega')}{\rho_F} \int d\Omega \alpha^2 F_{0,m}(\Omega) I(\omega, \Omega, \omega') \frac{\omega'}{|\omega'|} , \tag{14.69}$$

$$\delta\phi_n(\omega) = \sum_m \int d\omega' \frac{\rho(Z(0)\omega')}{\rho_F} \int d\Omega \alpha^2 F_{n,m}(\Omega) I(\omega, \Omega, \omega') \frac{\delta\phi_m(\omega')}{|\omega' Z(0)|} . \tag{14.70}$$

As we have mentioned, we can adopt any kind of basis function set f_n as long as it satisfies the ortho-normal condition, and we have already set f_0 to be the totally isotropic function, i.e., to be constant on each equal energy surface $\xi(\boldsymbol{\ell}) = \xi$. Now, as for f_2, we assume that it is proportional to $\delta\phi(\boldsymbol{\ell})$, which has d-symmetry and is of course orthogonal to f_0. (f_1 corresponds to p-symmetric function which is out of scope in the present book). Then (14.69)–(14.70) are written as

$$(1 - Z(0))\omega = \int d\omega' \frac{\rho(Z(0)\omega')}{\rho_F} \int d\Omega \alpha^2 F_{0,0}(\Omega) I(\omega, \Omega, \omega') \frac{\omega'}{|\omega'|}, \quad (14.71)$$

$$Z(0)\delta\Delta_2(\omega) = \int d\omega' \frac{\rho(Z(0)\omega')}{\rho_F} \int d\Omega \alpha^2 F_{2,2}(\Omega) I(\omega, \Omega, \omega') \frac{\delta\Delta_2(\omega')}{|\omega'|}. \quad (14.72)$$

Here the superconducting gap function Δ is defined as $\Delta = \phi/Z$. (Remark: This definition is valid to all temperature $T < T_c$, and it is known that the energy of a hole-carrier excitation is given as $E(\mathbf{k}) = \sqrt{\xi(\mathbf{k})^2 + \Delta(\mathbf{k})^2}$.)

Lastly, let us remember the two effects unique to the K–S model: (1) the electron–phonon coupling constant depends on spin, and (2) the electron–phonon interaction is ineffective for frequencies lower than ω_{hF}. The $\alpha^2 F_{0,0}$ in (14.71) comes from the exchange processes by the same hole carrier while $\alpha^2 F_{2,2}$ represents the effect of phonon exchange between carriers with different spins. To show the spin-dependencies, let us write the $\alpha^2 F$-terms explicitly with spin indices, which appear through ω_{hF}. As for the effect of ω_{hF}, we insert $\omega'^2/(\omega'^2 + \omega_{\mathrm{hF}}^2)$ in (14.71) and (14.72) instead of the simple cut-off which we introduced in the preceding chapter to avoid undesirable singular behaviour caused by the step function cut-off in the refined strong coupling treatment. Then the final form of the T_c-equation for the K–S model is

$$(1 - Z(0))\omega = \int d\omega' \frac{\omega'^2}{\omega'^2 + \omega_{\mathrm{hF}}^2} \frac{\rho(Z(0)\omega')}{\rho_F}$$
$$\times \int d\Omega \alpha^2 F_{\sigma,\sigma,0,0}(\Omega) I(\omega, \Omega, \omega') \frac{\omega'}{|\omega'|}, \quad (14.73)$$

$$Z(0)\delta\Delta_2(\omega) = \int d\omega' \frac{\omega'^2}{\omega'^2 + \omega_{\mathrm{hF}}^2} \frac{\rho(Z(0)\omega')}{\rho_F}$$
$$\times \int d\Omega \alpha^2 F_{\uparrow,\downarrow,2,2}(\Omega) I(\omega, \Omega, \omega') \frac{\delta\Delta_2(\omega')}{|\omega'|}. \quad (14.74)$$

14.6 Calculated Results of the Superconducting Transition Temperature and the Isotope Effects

14.6.1 Introduction

In solving (14.73) and (14.74), we make further approximations. First, even for high-T_c systems, T_c is low enough compared with typical phonon frequency such as the Debye frequency ω_D, i.e., $\omega_\mathrm{D} \gg T_\mathrm{c}$. Thus we neglect the Bose function $N(\Omega)$ which appeared in the expression of the I-function in (14.68). We also adopt the same approximation method as that treated by McMillan in [197]. Then, defining the non-dimensional coupling constant λ_d and non-dimensional self-energy interaction constant λ_0, respectively, by

$$\lambda_\mathrm{d} = \int d\Omega \frac{\alpha^2 F_{\uparrow,\downarrow:2,2}(\Omega)}{2\Omega} , \tag{14.75}$$

and

$$\lambda_0 = \int d\Omega \frac{\alpha^2 F_{\sigma,\sigma:0,0}(\Omega)}{2\Omega} , \tag{14.76}$$

we finally obtain the following equations to calculate T_c which we call the "T_c-equation":

$$1 = \lambda' \int_0^{\omega_\mathrm{D}} d\omega' \frac{\omega'^2}{\omega'^2 + \omega_\mathrm{hF}^2} \frac{[\rho(\omega' Z(0)) + \rho(-\omega' Z(0))]}{\rho_\mathrm{F}} \frac{\tanh(\omega'/2T_\mathrm{c})}{2\omega'} , \tag{14.77}$$

$$Z(0) = 1 + \lambda_0 \int_0^\infty \frac{\omega'^2}{\omega'^2 + \omega_\mathrm{hF}^2} \frac{[\rho(\omega' Z(0)) + \rho(-\omega' Z(0))]}{\rho_\mathrm{F}} \frac{\omega_\mathrm{D}}{(\omega' + \omega_\mathrm{D})^2} d\omega' , \tag{14.78}$$

where $\rho(\omega)$ is the energy-dependent density of states, ω_D the Debye frequency of the system and

$$\lambda' = \frac{\lambda_\mathrm{d}}{Z(0)} . \tag{14.79}$$

From the definition of Z-term by (14.78), $Z(0)$ is always larger than unity and therefore it is readily known that the effective coupling constant λ' becomes smaller than the "bare" coupling constant λ_d. In numerical calculations we adopt $\alpha^2 F(\Omega)$-functions calculated numerically by Ushio and Kamimura [112] for LSCO, which was given in Fig. 13.15 in Sect. 13.3. As for the spatial dependence of the gap function $\delta\Delta_2(\omega, \boldsymbol{k})$ in the $\boldsymbol{k}$-space, we have set

$$\delta\Delta(\omega, \boldsymbol{k}) = \delta\Delta_0(\omega) \cos[2\theta(\boldsymbol{k})] , \tag{14.80}$$

where $\theta(\boldsymbol{k})$ is an angle between $\boldsymbol{k}$ and the x-axis. This is equivalent to setting the function f_2, which appeared in the preceding section, as

$$f_2(\boldsymbol{k}) = \kappa_2 \cos[2\theta(\boldsymbol{k})] , \tag{14.81}$$

where κ_2 is a normalization factor. Then, reexpressing the $\alpha^2 F_{\uparrow,\downarrow:2,2}(\Omega)$ in (14.75) by $\alpha^2 F_{\uparrow\downarrow}^{(2)}(\Omega)$ in (13.23) in Chap. 13, $\alpha^2 F_{\uparrow\downarrow}^{(2)}(\Omega)$ in (14.75) is calculated as

14.6 Calculated Results of the Superconducting Transition Temperature

$$\alpha^2 F_{\uparrow\downarrow}^{(2)}(\Omega) = \int_{\xi(\boldsymbol{\ell})=0} \int_{\xi(\boldsymbol{k})=0} d\boldsymbol{\ell} d\boldsymbol{k} \alpha^2 F_{\uparrow\downarrow}(\Omega, \boldsymbol{\ell}, \boldsymbol{k}) f_2(\boldsymbol{\ell}) f_2(\boldsymbol{k}) \ . \quad (14.82)$$

Here we use the wave-vectors $\boldsymbol{\ell}$ and $\boldsymbol{k}$ instead of θ and θ' in (13.23) and neglect the ξ-dependence of the $\alpha^2 F$ since it is a slowly varying function of ξ, one-electron energy measured from the Fermi energy. As for the Debye frequency we put $\hbar\omega_D = 50$ meV, and for the non-dimensional coupling constant λ_d, we have obtained the value of $\lambda_d = 1.96$ [28, 112] in (13.24) in Chap. 13.

In the preceding chapter we treated ω_{hF} to be directly derived from the AF-correlation of the system, i.e., $\omega_{hF} = v_F/\ell_0$, where ℓ_0 is an average length of a metallic region, which is much longer than the spin-correlation length in the local AF order λ_s, due to the AF-fluctuation, as we described in detail in Fig. 11.2 in Chap. 11. In other words, if a "boundary" between two "distinct" AF-regions are static, then ℓ_0 becomes equal to λ_s, but in the present spin-fluctuating system in the 2D Heisenberg AF spin system, the fluctuation is dynamic, and so we do not have distinct boundaries. Itinerancy of a hole-carrier in the K–S model is acquired by taking an alternate multiplet structure of $^1A_{1g}$ and $^3B_{1g}$ from site to site without destroying the underlying AF-order. Since this physical picture is a quantum-mechanical one, in general a hole-carrier can itinerate over a distance much longer than λ_s itself. At the present, we do not have the relation between ℓ_0 and λ_s. Thus in the present numerical calculations, we treat the mean free path of a hole-carrier ℓ_0 to be a parameter which is proportional to the AF correlation length λ_s in its hole-concentration dependence.

14.6.2 The Hole-Concentration Dependence of T_c

We now show the results of numerical calculations. We have calculated the hole concentration (x)-dependencies of transition temperature $T_c(x)$ and isotope effect $\alpha(x)$ [26]. We first show the calculated results of $T_c(x)$ in Fig. 14.11. The existence of the lower cut-off parameter ω_{hF} described in Chap. 13 plays an important role together with the spin-dependent electron–phonon coupling. As we have seen, ω_{hF} is written as

$$\omega_{hF} = \frac{v(\boldsymbol{k})}{\ell_0(x)} \ , \quad (14.83)$$

where $v(\boldsymbol{k})$ denotes the group velocity of a hole-carrier with momentum $\boldsymbol{k}$ and $\ell_0(x)$, the "mean free path" of a hole-carrier in the metallic region at hole concentration x. Thus ω_{hF} has $\boldsymbol{k}$-dependence but in the present calculation we neglect it for simplicity. For $v(\boldsymbol{k})$, we adopt the averaged Fermi velocity $v_F(x)$ on the Fermi surface which corresponds to the hole concentration x and thus ω_{hF} has the x-dependence. Since we can easily show that $v_F(x) \gtrsim v(\boldsymbol{k})$ with $\boldsymbol{k}$ nearby the Fermi-surface which has a main contribution to the gap formation, this simplification has no effect of overestimating T_c. Now we have

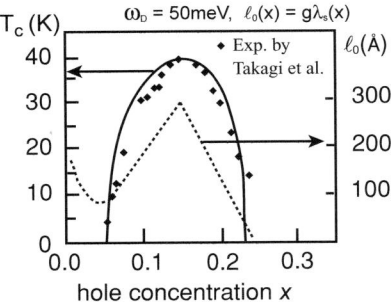

Fig. 14.11. The calculated x-dependence of T_c for LSCO. The *thick solid line* represents the calculated $T_c(x)$ curve and the *thin solid line* represents the x-dependence of "mean free length" ℓ_0 of hole-carriers. Experimental data are taken from [35]. $\ell_0(x)$ is fitted to reproduce the experimental results for $T_c(x)$, and x-dependence of ℓ_0 is taken so as to be consistent with neutron scattering experiments. For detail, see the text

$$\omega_{\mathrm{hF}}(x) = \frac{v_{\mathrm{F}}(x)}{\ell_0(x)} \; . \tag{14.84}$$

To calculate $T_c(x)$, we have to determine the x-dependence of $\ell_0(x)$, but up to the present we do not have a reliable method of calculating $\ell_0(x)$ from the K–S effective Hamiltonian directly. Then, as we have seen in Sect. 11.2 it is quite natural to assume that the "mean free path" of a hole-carrier ℓ_0 is proportional to the antiferromagnetic (AF) correlation length of the localized spin system, that is

$$\ell_0(x) = \gamma \lambda_s(x) \; , \tag{14.85}$$

where $\lambda_s(x)$ is the AF-correlation length at the hole concentration x. As regards the x-dependence of $\lambda_s(x)$, we use the experimental results of $\lambda_s(x)$ reported by Yamada et al. [143] and Lee et al. [53] up to the optimum doping ($x = 0.15$). Beyond $x = 0.15$, there are no available experimental data at present. Here we assume that $\lambda_s(x)$ decreases monotonically from $x = 0.15$ to $x_c = 0.25$, at which the AF order disappears, as we described in Sect. 12.5. For the very low hole-concentration region of $x < 0.05$ we adopt $\lambda_s = 3.8\,\text{Å}/\sqrt{x}$ following the experimental results by Birgeneau et al. [36, 145]. Using the above-mentioned x-dependence of $\lambda_s(x)$ and using a relation of $\ell_0 = \gamma \lambda_s(x)$, we have calculated T_c as a function of x. In doing so we have determined a value of constant γ to be 5, so as to reproduce the experimental value of $T_c = 40\,\text{K}$ at optimum doping $x_c = 0.15$ [35], by using the experimental value of $\lambda_s(x = 0.15) = 50\,\text{Å}$ [53, 143] and the density of states of the K–S model $\rho_{\mathrm{KS}}(\varepsilon)$ calculated in Sect. 11.5 (see Fig. 11.5). As seen from Fig. 14.11, we can successfully reproduce the observed bell shaped x-dependence of T_c. From the result we can say that the increase of T_c with increasing x in the underdoped region is due to the increase of the density of states with increasing the energy

14.6 Calculated Results of the Superconducting Transition Temperature

in $\rho_{\mathrm{KS}}(\varepsilon)$ while the decrease of T_c in the overdoped region with increasing x is due to the pair-breaking effect related to the decrease of the local AF order by the destruction of effective superexchange interactions in the AF-correlated region.

From the empirical value of $\gamma = 5$, we can say that a metallic region is spread over 25 nm for the optimum doping of LSCO. This means that the superconducting regions are inhomogeneous in cuprates, at least for LSCO. Recently Lang et al. [209] reported the granular structure of high-T_c superconductivity in underdoped Bi2212 by using scanning tunneling microscopy (STM). Their STM studies have revealed an apparent segregation of the electronic structure into superconducting domains that are $\sim$ 3 nm in size. Since STM studies concern with a much higher energy region over the Fermi energy, the discrepancy between the size of an empirically-determined metallic region and the size determined by STM might be due to the difference in the time scales caused from the different energy regions with which the present theory and STM experiments are concerned.

14.6.3 Isotope Effects

Next we show the calculated result of the isotope effect in Fig. 14.12 [26]. The isotope effect is measured by a constant α which is defined by

$$\alpha = -\frac{\mathrm{d}\ln T_\mathrm{c}}{\mathrm{d}\ln M}, \tag{14.86}$$

where M denotes the mass of constituent atoms. In the BCS theory, T_c is just proportional to the Debye frequency ω_D, and since ω_D generally has M-dependence of $\omega_\mathrm{D} \propto M^{-0.5}$, thus we have $\alpha = 0.5$ for weak coupled superconductors. But in the strong coupling regime, we do not have such

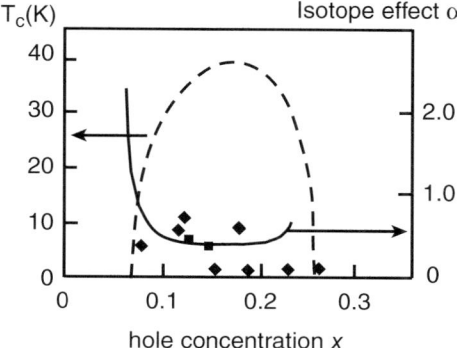

Fig. 14.12. The calculated x-dependence $\alpha(x)$ of the isotope effect. Here, the *solid line* represents $\alpha(x)$. The *dashed line* represents the calculated $T_\mathrm{c}(x)$, shown for reference

a simple relation, even in the conventional superconductors. In the present calculation we also use $\omega_D \propto M^{-0.5}$ with regard to the M-dependence of T_c. The calculated results are shown in Fig. 14.12. As seen from the figure, the isotope effect α becomes very large when x is small, while around the optimally doped region α is 0.39, which is slightly smaller than that of the BCS value, 0.5. In the present calculation, however, the mass of all constituent atoms has been changed by the ratio of the mass of ^{18}O to that of ^{16}O. Therefore, the calculated value of 0.39 is overestimated.

As we have seen, in T_c-equation (14.77), T_c depends on the density of states ρ through the factor $\rho = \rho(Z(0)\omega)$ where $Z(0) > 1$. Since the density of the states (DOS) of the K–S model $\rho_{KS}(\varepsilon)$ has a significant energy-dependence, the isotope effect becomes dependent on the hole-concentration x critically. In particular α becomes large in a low-concentration region where ρ_{KS} is small and thus T_c is also small. This peculiar x-dependence of isotope effect is one of the characteristic features of the isotope effect, obtained from the K–S model. In Fig. 14.2, the calculated results are compared with experimental results of LSCO by Crawford et al. [210, 211], who made the substitution of ^{18}O for ^{16}O in LSCO. In the figure the experimental results are shown by solid squares and diamonds. We can see that there are good agreement between the calculated results of the K–S model and the experimental results shown by the solid squares in the underdoped and overdoped region below 0.2 ($x < 0.2$), where the K–S model holds.

14.7 Final Remarks

Recently a number of papers related to the phonon mechanism have been published. Although the phonon mechanisms in these papers are different from the mechanisms of superconductivity due to the K–S model with two-component scenario described in Chap. 13 and the present chapter, in closing this chapter we would like to mention some of the papers as references [212, 213, 214, 215, 216, 217, 218, 219, 220].

References

1. H. Kamerlingh Onnes, Phys. Lab. Univ. Leiden 120b, 122b, 124c (1911).
2. J. Bardeen, L. N. Cooper and J. R. Schrieffer, Phys. Rev. 108, 1175 (1957).
3. J. G. Bednorz and K. A. Müller, Z. Phys. B64, 189 (1986).
4. M. K. Wu, J. R. Ashburn, C. J. Torng, P. H. Hor, R. L. Meng, L. Gao, Z. J. Huang, Y. Q. Wang, and C. W. Chu, Phys. Rev. Lett. 58, 908 (1987).
5. H. Maeda, Y. Tanaka, M. Fukutomi and T. Asano, Jpn. J. Appl. Phys. 27, L209 (1988).
6. A. Schilling, M. Cantoni, J. D. Guo and H. R. Ott, Nature 363, 56 (1993).
7. L. Gao, Z. J. Huang, R. L. Meng, J. G. Lin, F. Chen, L. Beauvais, Y. Y. Shen, Y. Y. Xue and C. W. Chu, Physica C213, 261 (1993).
8. L. Gao, Y. Y. Xue, F. Chen, Q. Xiong, R. L. Meng, D. Ramirez, and C. W. Chu, Phys. Rev. B50, 4260 (1994).
9. Y. Tokura et al, Phys. Rev. B39, 9704 (1989).
10. H. Takagi, T. Ido, S. Ishibashi, M. Uota, S. Uchida and Y. Tokura, Phys. Rev. B40, 2254 (1989).
11. G. Baskaran and P. W. Anderson, Phys. Rev. B37, 580 (1988).
12. P. W. Anderson, Science 235, 1196 (1987).
13. K. A. Müller, J. Superconductivity 12, 3 (1999), and related references therein.
14. H. Kamimura and T. Hamada, in *Int. Symp. on Physics in Local Lattice Distortions*, Vol. 554, eds. H. Oyanagi and A. Bianconi, pp. 3-37 (Melville, New York, 2001).
15. H. Kamimura and Y. Suwa, J. Phys. Soc. Jpn. 62, 3368 (1993).
16. E. Dagotto, Rev. Mod. Phys. 66, 763 (1994), and related references therein.
17. A. S. Alexandrov and N. F. Mott, *High Temperature Superconductivity and Other Superfluids* (Taylor & Francis, London, 1994).
18. P. W. Anderson, *The Theory of Superconductivity in the High-Tc Cuprates* (Princeton University Press, Princeton, 1997).
19. J. Ashkenazi, J. Superconductivity 10, 379 (1997).
20. A. S. Alexandrov, Int. J. Mod. Phys. B14, 3298 (2000), and related references therein.
21. D. Emin, in *Models and Methods of High Temperature Superconductivity*, Vol. 2, eds. J. K. Srivastava and S. M. Rao (Nova Science Publishers, New York, 2003).
22. *Models and Methods of High Temperature Superconductivity*, Vols. 1 and 2, eds. J. K. Srivastava and S. M. Rao (Nova Science Publishers, New York, 2003).
23. T. Yoshida, X. J. Zhou, T. Sasagawa, W. L. Yang, P. V. Bogdanov, A. Lanzara, Z. Hussain, T. Mizokawa, A. Fujimori, H. Eisaki, Z.-X. Shen, T. Kakeshita, and S. Uchida, (unpublished, 2004).

24. T. Yoshida, X. J. Zhou, M. Nakamura, S. A. Keller, P. V. Bogdanov, E. D. Lu, A. Lanzara, Z. Hussain, A. Ino, T. Mizokawa, A. Fujimori, H. Eisaki, Z.-X. Shen, T. Kakeshita, and S. Uchida, Phys. Rev. B63, 220501 (2001).
25. T. Yoshida, X. J. Zhou, T. Sasagawa, W. L. Yang, P. V. Bogdanov, A. Lanzara, Z. Hussain, T. Mizokawa, A. Fujimori, H. Eisaki, Z.-X. Shen, T. Kakeshita, and S. Uchida, Phys. Rev. Lett. 91, 027001 (2003).
26. H. Kamimura, T. Hamada, S. Matsuno and H. Ushio, J. Superconductivity 15, 379 (2002).
27. H. Kamimura, S. Matsuno, Y. Suwa and H. Ushio, Comments Cond. Mat. Phys. 7, 107 (1994).
28. S. Matsuno, H. Ushio, Y. Suwa and H. Kamimura, Int. J. Mod. Phys. B11, 3815 (1997).
29. H, Kamimura, T. Hamada, S. Matsuno and H. Ushio, Physica C364-365, 87 (2001).
30. H. Kamimura, S. Matsuno, Y. Suwa and H. Ushio, Phys. Rev. Lett. 77, 723 (1996).
31. Y. Tokura and T. Arima, Jpn. J. Appl. Phys. 29, 2388 (1990).
32. J. Akimitsu, S. Suzuki, M. Watanabe and H. Sawa, Jap. J. Appl. Phys. 27, L1859 (1988).
33. Y. Tokura, H. Takagi and S. Uchida, Nature 89, 345 (1989).
34. T. Ito, H. Takagi, S. Ishibashi, T. Ido and S. Uchida, Nature 350, 596 (1991).
35. H. Takagi, T. Ido, S. Ishibashi, M. Uota, S. Uchida and Y. Tokura, Phys. Rev. B40, 2254 (1989), related references therein.
36. G. Shirane, Y. Endoh, R. J. Birgeneau, M. A. Kastner, Y. Hidaka, M. Oda, M. Suzuki and T. Murakami, Phys. Rev. Lett. 59, 1613 (1987).
37. R. J. Birgeneau and G. Shirane, in *Physical Properties of High-Temperature Superconductors*, Vol. 1, ed. D. M. Ginsberg, p. 151 (World Scientific, 1989).
38. M. R. Presland, J. L. Tallon, R. G. Buckley, R. S. Liu and N. E. Flower, Physica C176, 95 (1991).
39. H. Yasuoka, T. Imai and T. Shimizu, in *Strong Correlation and Superconductivity*, Springer Series in Solid State Science Vol. 89, eds. H. Fukuyama, S. Maekawa and A. P. Malozemoff, p. 254 (Springer-Verlag, Berlin Heidelberg New York, 1989).
40. R. E. Walstedt and W. W. Warren, Jr., Science 248, 1082 (1990).
41. C. H. Pennington and C. P. Slichter, in *Physical Properties of High Temperature Superconductors* II, ed D. M. Ginsberg, p. 269 (World Scientific, Singapore, 1990).
42. J. Haase, P. Sushkov, P. Horsch, and G. V. M. Williams, Phys. Rev. B69, 094504 (2004), related refferences therein.
43. D. S. Marshall, D. S. Dessau, A. G. Loeser, C-H. Park, A. Y. Matsuura1, J. N. Eckstein, I. Bozovic, P. Fournier, A. Kapitulnik, W. E. Spicer, and Z.-X. Shen, Phys. Rev. Lett. 76, 4841 (1996).
44. A. Damascelli, Z. Hussain and Z.-X. Shen Rev. Mod. Phys. 75, 473 (2003), and related refferences therein.
45. M. R. Norman, H. Ding, M. Randeria, J. C. Campuzano, T. Yokoya, T. Takeuchi, T. Takahashi, T. Mochiku, K. Kadowaki, P. Guptasarma and D. G. Hinks, Nature 392, 157 (1998).
46. N. Miyakawa, P. Guptasarma, J. F. Zasadzinski, D. G. Hinks and K. E. Gray, Phys. Rev. Lett. 80, 157 (1998).

47. K. Yamanaka, A. Suzuki and M. Suzuki, J. Phys. Soc. Jpn. 71, 1405 (2002), and related references therein.
48. M. Oda, H. Matsuki and M. Ido, Solid State Commun. 74, 1321 (1990).
49. T. Matsuzaki, M. Ido, N. Momono, R. M. Dipasupil, T. Nagata, A. Sakai and M. Oda, Journal of Physics and Chemistry of Solids 62, 29 (2001), related references therein.
50. J. W. Loram, K. A. Mirza, J. R. Cooper, and W. Y. Liang, Phys. Rev. Lett. 71, 1740 (1993).
51. G. V. M. Williams, J. L. Tallon and J. W. Loram, Phys. Rev. B58, 15053 (1998).
52. M. Sato, S. Shamoto, K. Kiyokura, K. Kakurai, G. Shirane, B. J. Sternlieb and J. M. Tranquada, J. Phys. Soc. Jpn. 62, 263 (1993).
53. C.-H. Lee, K. Yamada, Y. Endoh, G. Shirane, R. J. Birgeneau, M. A. Kastner, M. Greven and Y-J. Kim, J. Phys. Soc. Jpn. 69, 1170 (2000), related references therein.
54. C. C. Homes, T. Timusk, R. Liang, D. A. Bonn, and W. N. Hardy, Phys. Rev. Lett. 71, 1645 (1993).
55. R. Hauff, S. Tajima, W.-J. Jang, and A. I. Rykov, Phys. Rev. Lett. 77, 4620 (1996).
56. Z.-X. Shen, D. S. Dessau, B. O. Wells, D. M. King, W. E. Spicer, A .J .Arko, D. Marshall, L. W. Lombardo, A. Kapitulnik, P. Dickinson, S. Doniach, J. DiCarlo, A. G. Loeser, C.-H. Park, Phys. Rev. Lett. 70, 1553 (1993).
57. H. Ding, M. R. Norman, J. C. Campuzano, M. Randeria, A. F. Bellman, T. Yokoya, T. Takahashi, T. Mochiku and K. Kadowaki, Phys. Rev. B54, R9678 (1996).
58. T. Imai, T. Shimizu, T. Tsuda, H. Yasuoka, T. Takabatake, Y. Nakazawa and M. Ishikawa, J. Phys. Soc. Jpn. 57, 1771 (1988).
59. D. J. Scalapino, Phys. Rep. 250, 329 (1995), related references therein.
60. P. G. de Gennes, *Superconductivity of Metals and Alloys* (Addison-Wesley Publishing Co., Inc. 1989).
61. Y. J. Uemura, J. Phys.:Condens. Matter 16, S4515-S4540 (2004), and related references therein.
62. W. N. Hardy, D. A. Bonn, D. C. Morgan, R. Liang, and K. Zhang, Phys. Rev. Lett. 70, 3999 (1993).
63. S. Hensen, G. Müller, C. T. Rieck and K. Scharnberg, Phys. Rev. B56, 6237 (1997).
64. K. A. Moler, D. J. Baar, J. S. Urbach, R. Liang, W. N. Hardy, and A. Kapitulnik, Phys. Rev. Lett. 73, 2744 (1994).
65. K. A. Moler, D. L. Sisson, J. S. Urbach, M. R. Beasley, A. Kapitulnik, D. J. Baar, R. Liang and W. N. Hardy, Phys. Rev. B55, 3954 (1997).
66. D. A. Wright, J. P. Emerson, B. F. Woodfield, J. E. Gordon, R. A. Fisher and N. E. Phillips, Phys. Rev. Lett. 82, 1550 (1999).
67. T. Staufer, R. Nemetshek, R. Hackl, P. Müller and H. Veith, Phys. Rev. Lett. 68, 1069 (1992).
68. T. P. Devereaux and D. Einzel, Phys. Rev. B51, 16336 (1995).
69. D. Shimada, Y. Shiina, A. Mottate, Y. Ohyagi and N. Tsuda, Phys. Rev. B51, 16495 (1995).
70. I. Iguchi, W, Wang, M. Yamazaki, Y. Tanaka and S. Kashiwara, Phys. Rev. B62, R6131 (2000), related references therein.

References

71. M. Sigrist and T. M. Rice, J. Phys. Soc. Jpn. 61, 4283 (1992).
72. D. A. Wollman, D. J. van Harlingen, W. C. Lee, D. M. Ginsberg and A. J. Leggett, Phys. Rev. Lett. 71, 2134 (1993).
73. J. R. Kirtley, C. C. Tsuei, J. Z. Sun, C. C. Chi, L. S. Yu-Jahnes, A. Gupta, M. Rupp and M. B. Ketchen, Nature 373, 225 (1995).
74. C. C. Tsuei and J. R. Kirtley, Rev. Mod. Phys. 72, 969 (2000), related references therein.
75. A. Sugimoto, T. Yamaguchi, I. Iguchi, Physica C367, 28 (2002).
76. C. C. Tsuei and J. R. Kirtley, Phys. Rev. Lett. 85, 182 (2000).
77. N.-C. Yeh, C.-T. Chen, G. Hammerl, J. Mannhart, A. Schmehl, C. W. Schneider, R. R. Schulz, S. Tajima, K. Yoshida, D. Garrigus, and M. Strasik, Phys. Rev. Lett. 87, 087003, (2001).
78. A. Jahn and E. Teller, Proc. Roy. Soc. A161, 220 (1937).
79. H. Aoki and H. Kamimura, Sollid State Commun. 63, 665 (1987).
80. K. H. Hock, N. Nickisch and H. Thomas, Helv. Phys. Acta 56, 237 (1983).
81. A. S. Alexandrov and N. F. Mott, Rep. Prog. Phys. 57, 1197 (1994), and related references therein.
82. H. Kamimura, Jpn. J. Appl. Phys. 26, L627 (1987).
83. H. Kamimura, Int. J. Modern Phys. B1, 873 (1987).
84. K. A. Müller, J. Superconductivity 12, 3 (1999).
85. K. A. Müller, Guo-meng Zhao, K. Conder and H. Keller, J. Phys. Condense Matter 10, L291 (1998).
86. D. Emin, in *Models and Methods of High-T_c Superconductivity*, Vol. 2, eds. J. K. Srivastava and S. M. Rao (Nova Science Publishers, New York, 2003), and related references therein.
87. A. S. Alexandrov and N. F. Mott, *Polarons and Bipolarons* (World Scientific, Singapore, 1995).
88. Y. Takada, Physica C364-365, 71 (2001).
89. G. Baskaran, Z. Zou and P. W. Anderson, Sold State Commun. 63, 973 (1987).
90. P. W. Anderson, Mater. Res. Bull. 8, 153 (1973).
91. S. Tajima, in *Advances in Superconductivity* III, eds. K. Kajiwara, and H. Hayakawa, p. 185 (Springer-Verlag, Tokyo, 1991).
92. S. Uchida, T. Ido, H. Takagi, T. Arima, Y. Tokura and S. Tajima, Phys. Rev. B43, 7942 (1991).
93. C. T. Chen, L. H. Tjeng, J. Kwo, H. L. Kao, P. Rudolf, F. Sette and R. M. Fleming, Phys. Rev. Lett. 68, 2543 (1992).
94. E. Pellegrin, N. Nücker, J. Fink, S. L. Molodtsov, A. Gutierrez, E. Navas, O. Strebel, Z. Hu, M. Domke, G. Kaindal, S. Uchida, Y. Nakamura, J. Markl, M. Klauda, G. Saemann-Ischenko, A. Krol, J. L. Peng, Z. Y. Li and R. L. Greene, Phys. Rev. B47, 3354 (1993).
95. V. J. Emery, Phys. Rev. Lett. 58, 2794 (1987).
96. F. C. Zhang and T. M. Rice, Phys. Rev. B37, 3759 (1988).
97. T. Tanamoto, H. Kohno, and H. Fukuyama, J. Phys. Soc. Jpn. 62, 717 (1993).
98. D. J. Scalapino, E. Loh and J. E. Hirsh, Phys. Rev. B34, 6694 (1987).
99. N. E. Bickers, D. J. Scalapino and S. R. White, Phys. Rev. Lett. 62, 961 (1989).
100. T. Moriya, in *Physics of High-Temperature Superconductors*, eds. S. Maekawa and M. Sato, p. 145 (Springer, 1992).

101. T. Takimoto and T. Moriya, J. Phys. Soc. Jpn 66, 1438 (1997), G. Esirgen and N. E. Bickers, Phys. Rev. B57, 5376 (1998). A number of studies using similar techniques were published.
102. A. J. Millis, H. Monien, and D. Pines, Phys. Rev. B42, 167 (1990).
103. M. Eto and H. Kamimura, J. Phys. Soc. Jpn. 60, 2311 (1991).
104. H. Kamimura and M. Eto, J. Phys. Soc. Jpn. 59, 3053 (1990).
105. K. Nomura and H. Kamimura, Solid State Commun. 111, 143 (1999).
106. N. Shima, K. Shiraisih, T. Nakayama, A. Oshiyama and H. Kamimura, Proc. 1st Int. Conf. on Elec. Materials in *New Materials and New Physical Phenomena for Electronics of the 21st Century*, eds. T. Sugano, R. P. H. Chang, H. Kamimura, I. Hayashi and T. Kamiya, p. 51 (JSAP-MRS, Pittsburg, 1988).
107. R. J. Cava, B. Batlogg, S. A. Sunshine, T. Siegrist, R. M. Fleming, K. Rabe, L. F. Schneemeyer, D. W. Murphy, R. B. van Dover, P. K. Gallagher, S. H. Glarum, S. Nakahara, R. C. Farrow, J. J. Krajewski, S. M. Zahurak, J. V. Waszczak, J. H. Marshall, P. Marsh, L. W. Rupp, Jr., W. F. Peck and E. A. Rietman, Physica C153-155, 560 (1988).
108. T. Egami, B. H. Toby, S. J. L. Billinge, Chr Janot, J. D. Jorgensen, D. G. Hinks, M. A. Subramanian, M. K. Crawford, W. E. Farneth and E. M. McCarron, in *High Temperature Superconductivity*, eds. J. Ashkenazi and G Vezzoli, p. 389 (Plenum Press, New York, 1992), and related references therein.
109. W. W. Shmahl, E. Salje and W. Y. Liang, Phil. Mag. Lett. 58, 173 (1988).
110. V. I. Anisimov, S. Yu. Ezhov and T. M. Rice, Phys. Rev. B55, 12829 (1997).
111. M. Merz, N. Nücker, P. Schweiss, S. Schuppler, C. T. Chen, V. Chakarian, J. Freeland, Y. U. Idzerda,M. Klaser, G. Müller-Vogt and Th. Wolf, Phys. Rev. Lett. 80, 5192 (1998).
112. H. Ushio and K. Kamimura, Int. J. Mod. Phys. B11, 3759 (1997).
113. H. Kamimura and H. Ushio, Solid State Commun. 91, 97 (1994).
114. A. Bussmann-Holder, K. A. Müller, R. Micnas, H. Büttner, A. Simon, A. R. Bishop, and T. Egami, J. Phys. Condensed Matter 13, L169 (2001), and related references therein.
115. K. Kuroki and H. Aoki, Phys. Rev. Lett. 69, 3820 (1992).
116. A. Bussmann-Holder, A. R. Bishop, H. Büttner, T. Egami, R. Micnas and K. A. Müller, J. Phys. Condensed Matter 13, L545 (2001).
117. H. Aoki and K. Kuroki, Phys. Rev. B 42, 2125 (1990).
118. W. Weber, A. L. Shelankov and X. Zotos, in *Mechanism of High Temperature Superconductivity*, Springer Series in Materials Science Vol. 11, eds. H. Kamimura and A. Oshiyama, p. 89 (Springer, Heidelberg, 1989), related references therein.
119. J. M. Tranquada, B. J. Sternlieb, J. D. Axe, Y. Nakamura and S. Uchida, Nature 375, 561 (1995).
120. Proceedings of the First International Conference on Stripe, Lattice Instability and High T_c Superconductivity, Special issue of J. Superconductivity Vol. 10, No. 4, eds A. Bianconi and N. L. Sani, (August 1997).
121. Proceedings of the Third International Conference on Stripe and High-T_c Superconductivity, Special issue of Int. J. Mod. Phys. B Vol. 14, Nos. 29, 30 & 31 (World Scientific, Singapore, 2000).
122. Proceedings of the international conference on Materials and Mechanisms of Superconductivity and High Temperature Superconductors VI: Physica C Vol. 341-348, pp. 1711-1804, eds. K. Salama, W-K Chu and P. C. W. Chu.

123. J. B. Boyce, F. Bridges, T. Claeson, T. H. Geballe, C. W. Chu and J. M. Tarascon, Phys. Rev. B35, 7203 (1987).
124. R. J. Cava, A. Santoro, D. W. Johnson, Jr. and W. W. Rhodes, Phys. Rev. B35, 6716 (1987). We use the lattice constants at 10K reported there.
125. M. A. Beno, L. Soderholm, D. W. Capone, II, D. G. Hinks, J. D. Jorgensen, J. D. Grace, Ivan K. Schuller, C. U. Segre and K. Zhang, Appl. Phys. Lett. 51, 57 (1987).
126. J. Kondo, in *Strong Correlation and Superconductivity*, Vol. 89, eds. H. Fukuyama, S. Maekawa and A. P. Malozemoff, p. 57 (Springer-Verlag, Heidelberg, 1989).
127. J. M. Tarascon, Y. Le Page, P. Barboux, B. G. Bagley, L. H. Greene, W. R. McKinnon, G. W. Hull, M. Giroud and D. M. Hwang, Phys. Rev. B37, 9382 (1988).
128. A. C. Wohl and G. Das, in *Modern Theoretical Chemistry*, Vol. 3, ed. H. F. Schaefer, p. 51 (Plenum Press, New York, 1977).
129. S. Kato and K. Morokuma, Chem. Phys. Lett. 65, 19 (1979).
130. M. Eto and H. Kamimura, Phys. Rev. Lett. 61, 2790 (1988).
131. H. Kamimura and A. Sano, Solid State Commun. 109, 543 (1999).
132. Y. Tobita and H. Kamimura, J. Phys. Soc. Jpn. 68, 2715 (1999).
133. B. Roos, A. Veillard and G. Vinot, Theoret. Chim. Acta. 20, 1 (1971).
134. T. H. Dunning Jr. and P. J. Hay, in *Modern Theoretical Chemistry*, Vol. 3, ed. H. F. Schaefer, p. 1 (Plenum, New York, 1977).
135. T. Shimizu, H. Yasuoka, T. Imai, T. Tsuda, T. Takabatake, Y. Nakazawa and M. Ishikawa, J. Phys. Soc. Jpn. 37, 2494 (1988).
136. H. Kamimura and A. Sano, J. Superconductivity 10, 279 (1997).
137. A. Sano, M. Eto and H. Kamimura, Int. J. Mod. Phys. B11, 3733 (1997).
138. H. L. Edward, A. L. Barr, J. T. Markert and A. L. de Lozanne, Phys. Rev. Lett. 73, 1154 (1994).
139. M. A. Mook, P. Dai, K. Salama, D. Lee, F. Dŏgan, G. Aepli, A. T. Boothroyd and M. E. Mostoller, Phys. Rev. Lett. 77, 370 (1996).
140. Sunhil Sinha, Talk at the Int. Conf. on Dynamical Energy Landscape and Functional Systems, March 29 to April 2, 2004, Santa Fe.
141. Y. Matsui, H. Maeda, Y. Tanaka and S. Horiuchi, Jpn. J. Appl. Phys. 27, L372 (1988).
142. T. E. Mason, A. Schroder, G. Aeppli, H. A. Mook and S. M. Hayden, Phys. Rev. Lett. 77, 1604 (1996).
143. K. Yamada, C. H. Lee, J. Wada, K. Kurahashi, H. Kimura, Y. Endoh, S. Hosoya, G. shirane, R. J. Birgeneau and M. A. Kastner, J. Superconductivity 10, 343 (1997).
144. A. Bianconi, P. Castrucci, A. Fabrizi, M. Pompa, A. M. Flank, P. Lagarde, H. Katayama-Yoshida and G. Calestani, Physica C162-164, 209 (1989).
145. R. J. Birgeneau, Y. Endoh, Y. Hidaka, K. Kakurai, M. A. Kastner, T. Murakami, G. Shirane, T. R. Thurston and K. Yamada, in *Mechanism of High Temperature Superconductivity*, Vol. 11, eds. H. Kamimura and A. Oshiyama, p. 120 (Springer-Verlag, Heidelberg, 1989).
See also, R. J. Birgeneau, D. R. Gabbe, H. P. Jenssen, M. A. Kastner, P. J. Picone, T. R. Thurston, G. Shirane, Y. Endoh, M. Sato, K. Yamada, Y. Hidaka, M. Oda, Y. Enomoto, M. Suzuki, and T. Murakami, Phys. Rev. B38, 6614 (1988).

146. K. B. Lyons, P. A. Fleury, J. P. Remeika, A. S. Cooper and T. J. Negran, Phys. Rev. B38, 6346 (1988).
147. L. F. Mattheiss, Phys. Rev. Lett. 58, 1028 (1987).
148. J. Yu, A. J. Freeman and J. H. Xu, Phys. Rev. Lett. 58, 1035 (1987).
149. P. Flude, *Electron Correlations in Molecules and Solids*, Vol. 100, (Springer-Verlag, Heidelberg, 1995).
150. C. Lanczos, J. Res. Natl. Bur. Stand. 45, 255 (1950).
151. T. Hamada, K. Ishida, H. Kamimura and Y. Suwa, J. Phys. Soc. Jpn. 70, 2033 (2001).
152. W. Weber, Phys. Rev. Lett. 58, 1371 (1987).
153. M. J. DeWeert, D. A. Papaconstantopoulos and W. E. Pickett, Phys. Rev. B39, 4235 (1989).
154. V. I. Anisimov, M. A. Korotin, J. Zaanen and O. K. Andersen, Phys. Rev. Lett. 68, 345 (1992).
155. J. C. Slater and G. F. Koster, Phys. Rev. 94, 1498 (1954).
156. K. Motizuki and N. Suzuki, in *Structural Phase Transition in Layered Transition-metal Compounds*, pp. 1-133 (Dordecht, Reidel, 1986), for the parameters needed for the calculation of the electron–phonon coupling of LSCO, see [112], references therein.
157. M. Shirai, N. Suzuki and K. Motizuki, J. Phys. Condens. Matter 2, 3553 (1990).
158. K. Motizuki, M. Shirai and N. Suzuki, in *Proceedings of Second CINVESTAV Superconductivity Symposium*, p. 176 (World Scientific, Singapore, 1990).
159. Y. Endoh et al., Physica C263, 394 (1996).
160. P. Aebi, J. Osterwalder, P. Schwaller, L. Schlapbach, M. Shimoda, T. Mochiku and K. Kadowaki, Phys. Rev. Lett. 72, 2757 (1994).
161. H. Krakauer, W. E. Pickett and R. F. Cohen, J. Supercond. 1, 111 (1988).
162. O. K. Anderson et al., Physica C185-189, 147 (1991).
163. A. Fujimori, JJAP Series 7 Mechanisms of Superconductivity, 125 (1992).
164. D. S. Desseau et al., Phys. Rev. Lett. 71 (1993) 2781.
165. D. M. King, Z.-X. Shen, D. S. Dessau, D. S. Marshall, C. H. Park, W. E. Spicer, J. L. Peng, Z. Y. Li, and R. L. Greene, Phys. Rev. Lett. 73, 3298 (1996).
166. P. Aebi, J. Osterwalder, P. Schwaller, L. Schlapbach, M. Shimoda, T. Mochida, K. Kadowaki, H. Berger and F. Levy, Physica C235-240, 949 (1994).
167. J. M. Luttinger and J. C. Ward, Phys. Rev. 118 (1960) 1417, J. M. Luttinger, Phys. Rev. 119 (1960) 1153.
168. T. E. Mason et al., Phys. Rev. Lett. 71 919 (1993).
169. S.-W. Cheong et al., Phys. Rev. Lett. 67 (1991) 1791.
170. Y. Endoh et al., JJAP Series 7 Mechanisms of Superconductivity, 174 (1992).
171. T. R. Thurston, P. M. Gehring, G. Shirane, R. J. Birgeneau, M. A. Kastner, Y. Endoh, M. Matsuda, K. Yamada, H. Kojima and I. Tanaka, Phys. Rev. B46, 9128 (1992).
172. Xiao-Gang Wen and P. A. Lee, Phys. Rev. Lett. 76, 503 (1996).
173. H. Ushio and H. Kamimura, J. Phys. Soc. Jpn. 64, 2585 (1995).
174. Nakamura and Uchida, Phys. Rev. B47, 8369 (1993).
175. H. Takagi, Y. Tokura and S. Uchida, in *Mechanisms of High Temperature Superconductivity*, Springer Series in Materials Science Vol. 11, eds. H. Kamimura and A. Oshiyama, p. 238 (Springer, Heidelberg, 1989).

176. T. Inoshita and H. Kamimura, Synthetic Metals, 3, 223 (1981).
177. J. M. Ziman, in *Electrons and Phonons*, (Oxford Univ. Press, London, 1960).
178. S. Mase et al., J. Phys. Soc. Jpn. 57 (1988) 607.
179. N. P. Ong, Z. Z. Wang, J. Clayhold, J. M. Tarascon, L. H. Greene and W. R. Mckinnon, Phys. Rev. B35 (1987) 8807.
180. T. Schimizu and H. Kamimura, J. Phys. Soc. Jpn. 59, 3691 (1990).
181. Y. Ando, Y. Kurita, S. Komiya, S. Ono, and K. Segawa, Phys. Rev. Lett. 92, 197001 (2004).
182. J. W. Loram, K. A. Mirza, J. R. Cooper and J. L. Tallon, J. Phys. Chem. Solids. 59, 2091 (1998).
183. H. Kamimura, T. Hamada and H. Ushio, Int. J. Mod. Phy. B14, 3501 (2000).
184. K. A. Mülller, Physica C341-348, 11 (2000).
185. R. Yoshizaki, N. Ishikawa, H. Sawada, E. Kits, and A. Tasaki, Physica C166, 417 (1990).
186. J. B. Torrance et al., Phys. Rev. B40, 8872 (1989).
187. D. C. Johnston, Phys. Rev. Lett 62, 957 (1989).
188. T. Nakano, M. Oda, C. Manabe, N. Momono, Y. Miura and M. Ido, Phys. Rev. B49, 16000 (1994).
189. M. Oda, T. Ohguro, N. Yamada, and M. Ido, Phys. Rev. B41, 2605 (1990).
190. S. Miyashita, J. Phys. Spc. Jpn. 57, 1934 (1988).
191. Y. Okabe and M. Kikuchi, J. Phys. Soc. Jpn. 57, 4351 (1988).
192. T. Nakano et al., J. Phys. Soc. Jpn. 67, 2622 (1998).
193. J. W. Loram et al., Physica C162, 498 (1990).
194. N. Suzuki and H. Kamimura, Solid State Commun. 8, 149 (1970).
195. Lanzara et al., Nature 412, 510 (2001).
196. M. Cardona, Physica C185-189, 65 (1991).
197. W. L. McMillan, Phys. Rev. 167, 331 (1968).
198. K. Motizuki and N. Suzuki, in *Structural Phase Transition in Layered Transition-metal Compounds*, eds. Dordecht, Reidel, pp. 1-133, (1986), for the parameters needed for the calculation of the electron–phonon coupling of LSCO, see [112], references therein.
199. H. A. Mook, M. Yethiraj, G. Aeppli, T. E. Mason and T. Armstrong, Phys. Rev. Lett. 70, 3490 (1993).
200. J. R. Schrieffer, *Theory of Superconductivity* (Perseus Books, 1999).
201. P. B. Allen and B. Mitrović, in *Solid State Physics*, Vol. 37, eds. H. Ehrenreich, F. Seitz and D. Turnbull, p. 1 (Academic Press, 1982).
202. A. B. Migdal, Zh. Exsp. Teor. Fiz. 38, 966 (1960), *Sov. Phys. – JETP(Engl. Transl.)* 11, 696 (1960).
203. G. M. Eliashberg, Zh. Eksp. Teor. Fiz. 38, 966 (1960), *Sov. Phys. – JETP(Engl. Transl.)* 11, 696, (1960).
204. L. P. Gorkov, Zh. Exsp. Teor. Fiz. 34, 735 (1958), *Sov. Phys. – JETP(Engl. Transl.)* 7, 505 (1958).
205. Y. Nambu, Phys. Rev. 117, 648 (1960).
206. P. B. Allen, Phys. Rev. B13, 1416 (1976).
207. D. J. Scalapino, J. R. Schriefer and J. W. Wilkins, Phys. Rev. 148, 263 (1966).
208. S. J. Nettel and H. Thomas, Solid State Commun. 21, 683 (1977).
209. K. M. Lang, V. Madhavan, J. E. Hoffman, E. W. Hudson, N. Eisaki, S. Uchida and J. C. Davis, Nature 415, 412 (2002).

210. M. K. Crawford, W. E. Farneth, E. M. McCarron, III, R. L. Harlow, A. H. Moudden, Science, 250, 1390 (1990).
211. M. K. Crawford, M. N. Kunchur, W. E. Farneth, E. M. McCarron III and S. J. Poon, Phys. Rev. B41, 282 (1990).
212. W. A. Little, K. Collins, and M. J. Holocomb, J. Superconductivity 12, 89 (1999).
213. S. R. Shenoy, V. Subrahmanyam and A. R. Bishop, Phys. Rev. Lett. 79, 4657 (1997).
214. G. Santi, T. Jalborg, M. Peter and M. Weger, Physica C259, 253 (1996).
215. J. C. Phillips, *Superconducting and Related oxides, Physics and Nanoengineering* III. Vol. 3481, eds. D. Pavana and Bozovic, p. 87 (1998).
216. D. Munzar and M. Cardona, Phys. Rev. Lett. 90, 077001 (2003).
217. S. Ishihara and N. Nagaosa, Phys. Rev. B69, 144520 (2004).
218. T. Egami, J. Superconductivity 15, 373 (2002).
219. Georgios Varelogiannis, Phys. Rev. B57, 13743 (1998).
220. M. Tachiki, M.Machida, and T. Egami, Phys. Rev. B67, 174506 (2003).

Subject Index

$Bi_2Sr_2Ca_{1-x}Dy_xCu_2O_{8+\delta}$, 113
$Bi_2Sr_2Ca_{n-1}Cu_nO_{4+2n+\delta}$, 11
$Bi_2Sr_2Ca_{n-1}Cu_nO_{4+2n}$, 10, 51
$Bi_2Sr_2CaCu_2O_{8+\delta}$, 5
$Bi_2Sr_{0.97}Pr_{0.03}CuO_{6+\delta}$ (Bi2201), 112
$Bi_2Sr_2CaCu_2O_{8+\delta}$ (Bi2212), 112
$Bi_2Sr_2CaCu_2O_{8+\delta}$, 51
Bi–Sr–Ca–Cu–O, 1
Bi–Sr–Ca–Cu–O materials, 51
Bi2201, 51
Bi2212, 5, 11, 30, 51, 53, 113
Bi2223, 11, 51
$Br_2Sr_2CaCuO_8$, 3
BSCCO, 11

crystal structures of $YBCO_{7-\delta}$, 43

energy bands in $YBa_2Cu_3O_7$, 95

Hg–Ba–Ca–Cu–O, 1
Hg1223, 53
$HgBa_2Ca_2Cu_3O_8$, 1
$HgBa_2CuO_4O_8$, 129
$HgBa_2Ca_2Cu_3O_{8+\delta}$ (Hg1223), 53

insulating $YBCO_6$, 43

La_2CuO_4, 1–3, 29, 108
$La_{1.94}Sr_{0.06}CuO_4$, 129

$La_{2-x}Sr_xCuO_4$, 4, 9, 33, 58, 111, 144
La, Ba_2CuO_4, 1
LSCO, 4, 9, 26, 85, 87, 108, 109, 111, 130, 131, 144, 151, 154, 188, 191

$Nd_{2-x-y}Sr_xCe_yCuO_4$, 9
$Nd_{2-x}Ce_xCuO_4$, 13
NSCCO, 9

superconducting $YBCO_7$, 43

$T_c = 35\,K$ to $TlBa_2Ca_2Cu_3O_9$, 1
$Tl_2Ba_2Ca_2Cu_3O_3$, 58
$Tl_2Ba_2CaCu_2O_8$, 58
Tl–Ba–Ca–Cu–O, 1
Tl1212, 53
$TlBa_2Ca_2Cu_3O_9$ ($HgBa_2Ca_2Cu_3O_8$, 3
$TlBa_2CaCu_2O_7$ (Tl1212), 53

undoped La_2CuO_4, 37

$YBa_2Cu_3O_6$, 43
$YBa_2Cu_3O_{6.91}$, 59
$YBa_2Cu_3O_7$, 3, 43, 93
$YBa_2Cu_3O_{7-\delta}$, 5
$YBa_2Cu_3O_{7-\eta}$, 1
YBCO, 14, 92
$YBCO_7$, 30, 53, 93, 95
$YBCOd_{7-\delta}$, 5

Index

$\alpha^2 F_{\uparrow\downarrow}(\Omega, \theta, \theta')$ 144–148, 151, 152
$\alpha^2 F_{\uparrow\downarrow}(\Omega, \boldsymbol{k}, \boldsymbol{k}')$ 139
$\alpha^2 F_{\uparrow\uparrow}(\Omega, \theta, \theta')$ 148–150
$\alpha^2 F_{\uparrow\uparrow}(\Omega, \boldsymbol{k}, \boldsymbol{k}')$ 139
Δ point 108

$^1A_1$ 31, 37, 44
$^1A_{1g}$ 30, 37, 39, 55, 189
$^1A_{1g}$ multiplets 55
$^1A_1$ multiplet 31
$^1A_{1g}$ multiplet 31, 39, 56, 109
$^1A_{1g}$ multiplet 84
a d hole 3
a hole-carrier motion across the boundary of a spin-correlated region λ_s 110
a single-electron-type band structure 107
a spin gap 131
a_1^* orbital 31
a_{1g} 29
A_{1g} phonon mode 144–146, 151
a_{1g}^* orbital 40, 55, 89, 115
AF Brillouin zone 111, 165
AF Heisenberg spin system 130
AF order 13, 68, 77, 81, 84
AF periodicity 164
AF reciprocal unit vector 114
AF-Brillouin zone (AF-BZ) 165
AF-correlation 13, 189, 190
AF-correlation length 172
AF-fluctuation 189
AF-localized spin system 161
AF-periodicity 165
AF-structure 12
AF-unit-vectors 164
alternant appearance of $^1A_{1g}$ and $^3B_{1g}$ multiplets 118

alternate appearance of a_{1g}^* and b_{1g} orbitals 117
angle resolved photoemission experiments 5
angle-resolved photoemission 112
angle-resolved photoemission spectroscopy (ARPES) 13, 139
angle-resolved-photoemission (ARPES) 17, 111
angular momentum components 85
anisotropic s-symmetry (extended s-symmetry) 14
anomalous behaviours of ρ_{ab} and ρ_c 122
anomalous concentration dependence of the electronic entropy 131
anomalous effective electron–electron interaction between holes with different spins 161
anomalous Green's function 179
anti-ferromagnetic spin fluctuation 24
anti-Jahn–Teller distortion 129
anti-Jahn–Teller effect V, 4, 5, 17, 25, 31, 33, 34, 39, 42, 57
antibonding a_{1g}^* orbital 30
antibonding b_{1g} orbital b_{1g}^* 40, 55
antibonding b_{1g}^* orbitals 84, 107
antiferromagnetic (AF) Brillouin zone 108
antiferromagnetic (AF) correlation length 190
antiferromagnetic (AF) insulator 2
antiferromagnetic (AF) order 55
antiferromagnetic (AF) ordering 66
antiferromagnetic unit cell in LSCO 127
apical O 15
apical O–Cu distance 44

Index

apical oxygen   4
apical oxygen site   44, 53, 59
Appearance of Repulsive Phonon-Exchange Interaction in the K–S Model   163
appearance of the d-wave superconductivity in LSCO system   159
Application of McMillan's Method to the K–S Model   183
Application of the Green's Function Method to a Superconducting State   179
APW band calculation   84
APW calculation   85
ARPES   13, 14, 24, 87, 109, 112, 113
ARPES experiment for $La_{2-x}Sr_xCuO_4$ 111
augmented-plane-wave (APW)   93
augmented-plane-wave (APW) calculation   85

$^3B_1$ 31, 44
$^3B_{1g}$ 30, 39, 189
$^3B_{1g}$ multiplet   31, 40, 55, 56, 84, 109
$^3B_1$ multiplet   31
b_{1g} 29
b_{1g} orbital   39, 89, 115
b_{1g} orbitals   55
b_{1g}^* 30
b_{1g}^* orbital   40, 89
Basic crystal structures of HTSC   10
basic elements of CuO_2-layer   15
basis function in the displaced structure   141
BCS   1, 23, 140, 161
BCS equation   169
BCS ground state   19
BCS pairing   162
Bednorz and Müller   3
Bi_2O_2 blocking layers   52
bipolaron   17
bipolarons   17
Bloch function in the displaced structure   141
Bloch functions in the K–S model   164
Bloch theorem   157
Bloch wave functions for up-spin and down-spin dopant holes   152

blocking layers   1, 9
bonding b_{1g} orbital   30
Bose condensation   17, 23
Bose Condensation Line of Holons   23
Bose condensation temperature   23
bubble diagram   176
by the pairing interaction   168
by transmission electron microscope   51

calculated x-dependence $\alpha(x)$ of the isotope effect   191
calculated x-dependence of T_c for LSCO   190
calculated electronic entropies of LSCO   128, 135
calculated entropy function based on the LDA band   129
calculated Fermi surface (FS) of LSCO   112
Calculated Results of the Superconducting Transition Temperature and the Isotope Effects   187
Calculation of the Spectral Functions for s-, p- and d-waves   143
CDW charge distribution   48
Change Density Wave (CDW)   45
character of wavefunctions   95
Choice of Basis Sets   38
cluster calculation   39
Cluster Models for Hole-Doped CuO_6 Octahedron   29
coexistence of the $^3B_{1g}$ and $^1A_{1g}$ multiplets   58
Coexistence of the $^1A_{1g}$ and the $^3B_{1g}$ Multiplets   58
coexistence of the spin-correlation of localized spins in the AF order and superconductivity   59
coexistence of two orbitals a_{1g}^* and b_{1g} 78
coherent length of Cooper pair   81
coherent motion of a dopant hole   57
conduction mechanism along the c-axis   123
conservation of pseudo-momentum   164

contribution for the d-wave components of the spectral function, $\alpha^2 F^{(2)}_{\uparrow\downarrow}(\Omega)$ 151
Cooper pair coupling formation 171
Cooper pairs of $d_{x^2-y^2}$ symmetry 139
Cooper-pairing 162
coulomb repulsion 16, 181
coulomb repulsion between hole-carriers 81
cross-over phenomena 114
crystal structures of $Bi_2Sr_2Ca_{n-1}Cu_nO_{4+2n+\delta}$ with $n = 1, 2, 3$ 10
Cu $d_{x^2-y^2}$ orbitals 21
Cu $d_{x^2-y^2}$ and d_{z^2} orbitals 29
Cu house 62
Cu localized spins 61
Cu_2O_{11} 37
Cu–apical O distance 17, 31, 33, 53
Cu–apical O(2) distance 34
Cu–O chain 44, 45, 48
Cu-chain 43
CuO_2 1
CuO_2-layer 3, 15
CuO_2-layers 9
CuO_2 planes 2
CuO_4 plane 15
CuO_4 square 10
CuO_5 3
CuO_5 pyramid 4, 15, 25, 30, 44, 45, 51, 52
CuO_5 pyramids 9, 15
CuO_6 3
CuO_6 octahedron 3, 4, 15, 25
CuO_6 octahedron, CuO_5 pyramid 10
CuO_6 octahedrons 9, 15
CuO_6 octahedron 29
CuO_6 Octahedron Embedded in LSCO 37
cuprate family 10
cuprates 1
cut-off parameter 189

1D chain 68, 72
1D chain lattice 63, 64
1D chain system 73
2D AF Heisenberg spin system 129
2D AF order 80
2D Heisenberg AF spin system 189

2D square 64
2D square lattice 63, 67, 71
2D square system 72, 73
$d_{x^2-y^2}$ 38
$d_{x^2-y^2}$ orbital 3
d_{z^2} 38
d_{z^2} orbital 27, 31
$d_{x^2-y^2}$-gap 167
$d_{x^2-y^2}$-gap symmetry 171
$d_{x^2-y^2}$-symmetry pairing 167
$d_{x^2-y^2}$-wave pairing 167
$d_{x^2-y^2}$-symmetry 14
$d_{x^2-y^2}$-wave superconductivity 24
d–p model 21, 22
d-symmetry 169, 170, 187
d-wave 13, 170
d-wave component $\alpha^2 F^{(2)}_{\uparrow\downarrow}(\Omega)$ 150
d-wave component in the spectral function 151
d-wave component of the spectral function, $\alpha^2 F^{(2)}_{\uparrow\downarrow}(\Omega)$ 153
d-wave pairing in the phonon mechanism 154
d-wave symmetry 6
Debye frequency 188
density of states 5, 117, 128, 134, 188, 190
density of states for the highest occupied band in LSCO 128
density of states of hole carriers 140
derivatives of the Slater Koster parameter 143
derivatives of transfer integrals 143
difference of one-electron level Δ between O $2p_\sigma$ and Cu $d_{x^2-y^2}$ 21
diffuse X-ray scattering 45
Doping dependence of Fermi surfaces of $La_{2-x}Sr_xCuO_4$ 112
double zeta 38
dynamical modulation of spin and charge in LSCO 27
dynamical stripe 27

E_u phonon 148
E_u phonon mode 146–150, 152
effect of Coulomb repulsion between hole-carriers 181

effective electron–electron interaction   162
effective Hubbard Hamiltonian   23
effective mass   137
effective one-electron-type band structure   83, 107
effects of the exchange integral   83
electrical resistivity   5, 131
electrical resistivity of $La_{2-x}Sr_xCuO_4$ 122
electron configurations   39
electron–phonon coupling constant   17, 176, 187
electron–phonon coupling constant for d-wave pairing, λ_d 151
electron–phonon coupling constants   164
Electron–Phonon Coupling Constants for the Phonon Modes in LSCO   139
electron–phonon interaction   139, 140, 143, 162, 163, 166, 171, 173, 178
electron–phonon interaction matrix element   140, 152, 157
electron–phonon scattering between the neighbouring FS pockets   124
electron–phonon spectral functions   6, 139, 144
electron-correlation   15
electron-correlation effect   38
electron-coupling constants of different spin carriers   165
electron-doped materials   14
electron-doped system   12
electron-lattice interaction   6
electronic entropy   5, 127, 128, 131
Electronic Structure of a Hole-Doped CuO_5 Pyramid   30
electronic structure of cuprates   15
electronic structure of LSCO   85
electronic structure of the K–S model   109
electronic structures   3
Electronic Structures of a Hole-Doped CuO_6 Octahedron   30
"elongated" and "deformed" octahedrons   17
Emery   21

energy bands   5
energy difference   41, 44, 52
Energy Difference between $^1A_1$ and $^1B_1$ Multiplets   44
energy difference between the $^3B_{1g}$ and $^1A_{1g}$ multiplets   91
energy difference between the $^1A_1$ and $^3B_1$ states   53
energy difference between the $^3B_{1g}$ and the $^1A_{1g}$ multiplets   42
energy difference between the $^3B_1$ and the $^1A_1$ multiplets   45
entropy   133
Exact Diagonalization Method   63
excess oxygen   51
exchange interaction between the spin of a dopant hole and a localized spin   91
exchange interaction between the spins of a dopant hole and a localized spin   83
exchange interaction between the spins of hole-carriers and of the localized holes in the same CuO_6 octahedron   26
exchange interactions between the spins of dopant holes and $d_{x^2-y^2}$ localized holes within the same CuO_6 octahedron   59
existence of static anti-ferromagnetic order   171
Existence of the Antiferromagnetic Spin Correlation   58
Experimental Evidence in Support of the K–S Model   58
Explicit Forms of the Electron–Phonon Interaction   154
explicit forms of the electron–phonon interaction $g_\mu^\alpha(\bm{k}\bm{k}')$ 154
expression of a spectral function in the tight binding form   141
extended s-wave   170

failure of the application of one-electron band theory   3
Fermi arcs   17, 111
Fermi liquid   22
Fermi liquid picture   15, 131

Fermi surface   17, 110, 111, 126, 127, 173
Fermi surface structure   110, 123
Fermi surfaces   5, 107, 122, 141
Fermi-velocity   109, 173
Fermion condensation   23
Feynman equation   178
Feynman graph expansion   175
Feynman–Dyson equation   175, 180
finite lifetime due to the dynamical 2D-AF fluctuation   171
finite lifetime effect   172
finite size of a metallic region   111
finite size of the AF-correlation length   161
finite spin-correlation length   161
finite-size effect of a metallic state   171
finiteness of AF-ordered range   172
first-principles band calculation   85
first-principles calculation   33, 83
first-principles calculations   4
first-principles cluster calculation   75
first-principles cluster calculations   25
first-principles norm-conserving pseudopotential method   32
flat band   109, 127
flat band of the heavy effective mass   130
formation of Cooper pairs   166
formation of the Cooper-pair   140
formula for the Hall coefficient   125
free energies   132
Free Energies of the SF- and LF-Phases   132

G_1 point   108, 113
gap   173
gap function   158, 187, 188
gap function of the superconductor   180
Gaussian functions   38
George Bednorz and Karl Alex Müller   1
granular structure of high-T_c superconductivity   191
graphite intercalation compounds (GICs)   123
Green's function method   174

Green's Function Method in the Normal State   174

half width of the anti-ferromagnetic incommensurate peaks   172
half-filled metals   16
Hall angle   127
Hall coefficient of LSCO   125
Hall effect   5, 125
Heisenberg spin Hamiltonian   75
Heitler–London states   37
high temperature superconductivity (HTSC)   1
"high-energy" and "low-energy" pseudogaps   131
"high-energy" pseudogap   136
high-energy pseudogap   13, 114, 121
historical development of critical temperatures   2
hole-concentration dependence of the spin correlation length   58
hole-concentration dependences of the spin susceptibility   130
hole-doped CuO_2-layer   19
Hole-Doped CuO_5 Pyramid in $YBa_2Cu_3O_{7-\delta}$ 43
Hole-Doped CuO_6 Octahedrons in LSCO   39
hole-doped system   12
holon   19, 20, 23
HTSC   15
Hubbard   16
Hubbard bands   108
Hubbard interaction   21
Hubbard-U interaction   81, 84
Hund's coupling   44, 53, 55, 61
Hund's coupling energy   91
Hund's coupling exchange constant   60
Hund's coupling spin-triplet   5
Hund's coupling spin-triplet multiplet   5
Hund's coupling triplet   26, 30, 31, 40, 56, 58, 61, 121

important role of strong electron correlation   3
in-plane O $2p_\sigma$ orbitals   21
in-plane O   15

in the LF-phase   134
in the SF-phase   134
in-plane modes   151
in-plane-resistivity   122
inadequacy of the electronic band structure based on the LDA band   129
incommensurability   114, 115
incommensurate peaks   114
incommensurate peaks in the spin excitation spectra of LSCO   114
inhomogeneous   27
inhomogeneous distribution of metallic regions   123
inhomogeneous hole distribution in the Cu–O chain   44
inhomogeneous superconducting state   26
inter-pocket scattering   123
interplay between the electron correlation and the local lattice distortion   137
intra-pocket scattering   123
Intuitive illustration of the averaging of the pairing interaction   168
irreducible self-energy   177
irreducible self-energy part   175
irreducible vertex   177
irreducible vertex part   177
isotope effect   129, 161
isotope effects   191

Jahn–Teller (JT) polarons   17
Jahn–Teller distortion   3, 17, 29
Jahn–Teller effect   4, 25, 29
Jahn–Teller interaction   3
Jahn–Teller theorem   17
Jahn–Teller-effect   31
JT-polarons   17

K_2NiF_4 137
K–S entropy function   129
K–S Hamiltonian   59, 63, 83, 107
K–S model   4, 25, 26, 63, 64, 68, 73, 74, 113, 114, 119, 122, 123, 126, 127, 129, 152, 161, 171, 189, 190, 192
Kamerlingh Onnes   1

Kamimura–Suwa (K–S) Model: Electronic Structure of Underdoped Cuprates   55
Kamimura–Suwa model   25, 26, 55

L edges   58
L_3 line   58
Lanczos method   63, 65
large Fermi surface   113, 132
large FS   130
large polaron   4, 17
large polarons   137
LDA + U band calculation   26
LDA band   110
LDA density of states   128
LDA entropy function   129
LDA+U band calculations   24
LDA+U 85
length of a metallic region   189
length of a wider metallic region   109
LF-phase   132
LF-phases   136
life time of carriers   111
lifetime broadening effects   113
lifetime effect   171
Ligand Field Theory   29
linear temperature-dependence of the resistivity   124
linear temperature-dependent metallic conductivity   122
linearized Eliashberg equation   158
linearized Feynman–Dyson equation   182
linearized-augmented-plane-wave (LAPW) band   93
local AF order   6, 26, 57, 109, 130, 161, 189
local antiferromagnetic (AF) order   55, 107, 139
local antiferromagnetic ordering   133
local density approximation (LDA)   110, 133
local density functional formalism (LDF)   32
localized spin system   171
localized spin-polaron   78
localized spins   55, 64, 83
long range Néel order   109
low-energy pseudogap   13, 137

lower crossover line   131
LSCO   139, 153
Luttinger's theorem   114

$\overline{M}$ point   109, 113
Madelung potential   34, 44, 47, 48, 52, 53, 91
magnetic excitation spectrum   131
Magnetic Properties   130
magnetic susceptibility   131
many-body effects   26
many-body interactions   84
many-body-effect including energy band   26
Many-Electron Energy Bands   87
many-electron states   38
many-electron wave function   116
Many-Electrons Band Structures   107
mass enhancement of the hole-carriers   129
material dependence of the maximum T_c 11
McMillan's equation   185
McMillan's method   183
McMillan's method to the T_c-equation   181
McMillan's treatment   181
MCSCF-CI   5
MCSCF-CI calculations   48
MCSCF-CI cluster calculations   60
MCSCF-CI method   37, 44, 51
mean field approximation   23
"mean free path" of a hole-carrier   190
mean-field approximation   83, 107, 121
mean-field approximation for the K–S Hamiltonian   83
mean-field calculation   23
mean-field treatment   26
mean-free path of carriers   113
Mechanism of High Temperature Superconductivity   161
mid-gap state   21
Migdal's approximation   178
Models of High-Temperature Superconducting Cuprates   15
modified density of states for the b_{1g}^* band   134
modified LDA state   134, 135

"molecular field" of the localized spins   84
momentum-dependent spectral function   141, 144–152
Mott–Hubbard insulator   108
Multi-Configuration Self-Consistent Field Method with Configuration Interaction (MCSCF-CI)   25
multi-configuration self-consistent field method with configuration interaction   5, 37
"multi-layered materials   11
multiplet   29
multiplets   5

Néel order   12, 66
Néel temperature   2, 130
nesting vectors   114
nesting wave vectors   114
neutron inelastic scattering experiments   109
neutron scattering   58
neutron scattering experiment   114
neutron time of flight experiments   25
NMR relaxation rate experiments   13
non-dimensional coupling constant   188
non-dimensional self-energy interaction constant   188
"Normal"-process   166
Normal modes   154
normal periodicity   165
normal state properties   121

O $2p$ orbitals   38
Observed doping dependence of Fermi surfaces of $La_{2-x}Sr_xCuO_4$ 113
Occurrence of the d-wave Symmetry   167
octahedral symmetry   29
off-diagonal orbital correlation function   74, 79
off-diagonal orbital correlation functions   76, 77
on-site Coulomb repulsion   16
on-site Hubbard interaction   18
on-site energies due to the displacement of atoms   143
on-site energy   142

on-site energy of $^3B_{1g}$ and $^1A_{1g}$ 144
one-body Green's function of a carrier system   174
one-electron-type Hamiltonian   83
one-orbital case   77
optimized Cu–apical O distance   33
optimum doping   13, 130
Orbital Correlation Functions   74
ordinary Brillouin zone   127
origin of the T^2 dependence of $\cot\theta_H$ 130
Origin of the High-Energy Pseudogap   131, 135
out-of-plane resistivity   122
overdoped   13
overdoped region   130

p_σ orbital   3
$2p_\sigma$ orbitals   21
pair-breaking effect   191
parameters in the K–S Hamiltonian   60
Pauli-like behaviour   114, 130
penetration depth measurement   13
perturbation expansion   22
phase diagram   12
Phase Diagram of Cuprates   12
Phase diagram of the t–J model   23
Phillip Anderson   3
phonon Green's function   175
π-junction experiment   14
point charges   34
polarization-dependent X-ray absorption measurements   58
polarized X-ray absorption spectra   26
polarized X-ray absorption spectra (XAS)   116
potential   44
pseudo-momentum conservation   165
pseudogap   131
pseudogap transition   23
Pyramid in $Bi_2Sr_2CaCu_2O_{8+\delta}$ 51

quasi-particle excitation   23
Quasi-particle Excitations   18

radial distribution function   79, 80
Radial Distribution Function for Two Hole-Carriers   79
radial distribution functions   81
Raman scattering   13
reciprocal lattice   115
reciprocal lattice vector in the AF Brillouin zone   143
renormalization factor   180
renormalization function   158
renormalized factor   186
Repulsive Electron–Phonon Interaction between Up- and Down-Spin Carriers with Different Wave Function   156
Resistivity   122
resistivity formula due to collisions of the hole-carriers with lattice phonons   123
Resonating Valence Bond (RVB)   18
resonating valence bond (RVB) state   18
Rice   22
RVB state   18, 19

s-symmetry   170
s-wave   13, 170
s-wave component of spectral function $\alpha^2 F^{(0)}_{\uparrow\downarrow}(\Omega)$ 150
s-wave symmetry   181
saddle point of the van Hove singularity   109
saddle point singularity   128
saddle-point type   117
scanning tunneling microscopy (STM)   45
Schematic picture for the selection rule   166
Schematic picture of scattering   168
Schematic view of the K–S model in LSCO   56
Schematic view of the K–S model in $YBCO_7$ and Bi2212   57
second order perturbation theory   167
selection rule   163
SF-phase   132
SF-phases   136
Simple Folding of the Fermi Surface into the AF Brillouin Zone   118
single doped hole-carrier   64
single Hubbard model   18
Single Orbital State   76

single-band Hubbard Hamiltonian 18
single-electron-type band structure 26
singlet pairs 18
site-specific X-ray absorption spectroscopy 26, 59
SK parameters 87
Slater–Koster (SK) parameters 84
Slater–Koster fits for $YBa_2Cu_3O_7$ 93
Slater–Koster method 85
slave-boson 23
slave-fermion 23
small Fermi surface 113
small Fermi surfaces 113, 132
small polaron 17
specific heat 13
spectral function 140
spectral function, which contributes to mass enhancement 153
spin fluctuation 24
spin fluctuation effect 67
spin fluctuation effect in a 1D system 72
Spin Fluctuation Models 24
spin frustration effect in the AF order 68
spin susceptibility 114
spin susceptibility $\chi(q,\omega)$ 24
spin-correlated region 5, 27, 55, 61, 83, 109
spin-correlated regions 129
spin-correlation function 68–73, 78, 80
Spin-Correlation Functions 66
spin-correlation length 6, 55, 56, 58, 109, 171, 189
spin-excitation-energy in the metallic region of a finite size 137
spin-fluctuating system 189
spin-fluctuation 6
spin-fluctuation effect 27
Spin-Gap State 23
spin-wave excitation 137
spinon 20, 23
spinon pairing condensation 23
Spinon-pair Condensation Line 23
static antiferromagnetic order 16
STM 191
Stripe Phase 27
strong U effect 107

Strong Coupling Treatment of Conventional Superconducting System 174
strong electron–phonon interaction 161
strong electron-lattice interactions 3
strong quantum fluctuation effect 18
strongly correlated interaction 22
substitution of ^{18}O for ^{16}O in LSCO 192
superconducting transition 23
superconductivity 12
superexchange interaction 16, 18, 83, 84, 107, 109
superexchange interaction between the Cu $d_{x^2-y^2}$ localized spins 59
superexchange interactions 64
Suppression of Superconductivity by Finiteness of the Anti-Ferromagnetic Correlation Length 171
Suppression of Superconductivity Due to the Finite Lifetime Effect 173
Symmetry of the Gap 13

T-phase 1, 9, 10
T'-phase 1, 9, 10
T^*-phase 1, 9, 10
T_c-Equation in the Strong Coupling Model 181
t–J model 22, 30, 79
t–J Hamiltonian 23
T-linear temperature dependence 124
temperature dependence of the Hall effect 127
temperature dependence of the in-plane resistivity 123
temperature- and concentration-dependence of the c-axis conduction 123
tetragonal symmetry 29
the b_{1g} bonding orbital 26
The calculated concentration dependence of the Hall coefficient 126
The calculated temperature dependence of the resistivity in the ab plane, ρ_{ab}, of LSCO for 124

the electron–phonon mediated
    superconductivity 85
The Hole-Concentration Dependence of
    T_c 189
the K–S model 128
the Kamimura–Suwa model 4
The mixing ratio of the in-plane Op_σ,
    apical Op_z, Cu $d_{x^2-y^2}$ and Cu d_{z^2}
    orbitals 116
the sign change of the effective
    interaction 171
Theories for HTSC 17
three-dimensional AF ordering 2
tight binding (TB) form 117
Tight Binding (TB) Functional Form
    115
tight binding (TB) Hamiltonian 85
tight binding Hamiltonian 141
tight binding model 143
transfer integral 16, 21
transfer interaction 142
transfer interaction between neighbouring CuO_6 octahedrons
    59
transfer interactions 84
transition temperature 161, 169, 173
transition-lines 13
traveling time of a hole-carrier 109
two dimensional AF Brillouin zone
    112
two doped hole-carriers 64
two story house model 61
two-component mechanisms 25
two-component scenario 118
two-component system 33
two-component theory 27

two-dimensional (2D) antiferromagnetic
    (AF) behaviour 130
two-dimensional AF Heisenberg spin
    system 27

"Umklapp-process 166
underdoped regime 122
underdoped region 13, 130
upper crossover line 131
upper Hubbard b_{1g}^* band 108

Validity of the K–S Model in the
    Overdoped Region 130
van Hove singularity 117, 150, 167
variational method 37
variational trial function 37
velocity of a hole-carrier 173
virtual crystal approximation 32
virtual crystal structure 52
virtual phonon exchange process 173

wave functions of an up-spin carrier
    115
Wavefunctions of a Hole-Carrier 115
weak coupling-equation 167

x-dependences of entropy 136

Zhang 22
Zhang–Rice exchange constant 60
Zhang–Rice singlet 22, 26, 30, 31, 37,
    39, 55, 56, 58, 61, 63, 76, 79, 121
Zhang–Rice spin-singlet 5
Zhang–Rice spin-singlet multiplet 5
"zigzag" like state 66
"zigzag"-like states 75
zigzag-like behaviour 81